The Mexican
Profit-Sharing
Decision

The Mexican Profit-Sharing Decision

Politics in an Authoritarian Regime

Susan Kaufman Purcell

University of California Press
Berkeley   Los Angeles   London

University of California Press
Berkeley and Los Angeles, California

University of California Press, Ltd.
London, England

To John

Contents

Acknowledgments

This study would not have been possible without the help and cooperation of my many kind Mexican friends and acquaintances. Although I cannot thank them each here, I have listed most of their names in the Appendix. I would, however, specifically like to thank Luis Macías Cardone and Alfonso Ogarrio for the very special effort they made in my behalf during my stay in Mexico.

I am also greatly indebted to several of my former professors at Columbia University. My greatest debt is to Douglas A. Chalmers, who worked closely with me on all versions of the manuscript and whose ideas about Latin American politics have had a profound influence on my thinking. I am also grateful to Dankwart A. Rustow, whose vast and varied knowledge was of immeasurable value. In addition, I would like to thank Ronald M. Schneider for his substantial contribution. More recently, I have benefited from the excellent and constructive comments of Professors Charles W. Anderson of the University of Wisconsin, Martin C. Needler of the University of New Mexico, William S. Tuohy of the University of California, Davis, and James W. Wilkie of the University of California, Los Angeles, and from the editorial advice of Marya W. Holcombe.

ACKNOWLEDGMENTS

I would also like to express my appreciation to the United States Department of Health, Education and Welfare for a Fulbright-Hays Fellowship that enabled me to undertake field work in Mexico and for a National Defense Foreign Language Fellowship that supported me during the year in which I wrote the first draft of this manuscript. I am also grateful to Professor Charles Wagley, former head of the Institute of Latin American Studies of Columbia University, for granting me a Latin American Institute Traveling Fellowship that enabled me to extend my stay in Mexico.

Two other people made contributions of utmost importance. I would like to thank my mother, Ilene Kaufman, for her support. And to my husband and colleague, John F. H. Purcell, I would like to express my gratitude, for his substantive comments and ideas, his patience, good humor and love, and his faith and encouragement.

S.K.P.

List of Abbreviations

ABM	Asociación de Banqueros Mexicanos
	Mexican Bankers' Association
AMIS	Asociación Mexicana de las Instituciones
	de Seguridad
	Mexican Association of Insurance Institutions
BUO	Bloque de Unidad Obrera
	Labor Unity Bloc
CCI	Confederación de Campesinos Independientes
	Independent Peasants' Confederation
CGT	Confederación General de Trabajadores
	General Workers' Confederation
CNC	Confederación Nacional Campesina
	National Peasants' Confederation
CNCRM	Confederación Nacional de Cooperativas de la
	República Mexicana
	National Confederation of Cooperatives of
	the Mexican Republic

CNOP	Confederación Nacional de Organizaciones Populares
	National Confederation of Popular Organizations
CNPP	Confederación Nacional de Pequeños Proprietarios Agrícolas, Ganaderos y Forestales
	National Confederation of Small Farm, Ranch and Forest Owners
CNT	Confederación Nacional de Trabajadores
	National Workers' Confederation
CONCAMIN	Confederación de Cámaras de Industria
	Confederation of Chambers of Industry
CONCANACO	Confederación de Cámaras Nacionales de Comercio
	Confederation of National Chambers of Commerce
COPARMEX	Confederación Patronal de la República Mexicana
	Employers' Confederation of the Mexican Republic
CROC	Confederación Revolucionaria de Obreros y Campesinos
	Revolutionary Confederation of Workers and Peasants
CROM	Confederación Regional de Obreros Mexicanos
	Regional Confederation of Mexican Workers
CRT	Confederación Revolucionaria de Trabajadores
	Revolutionary Confederation of Workers
CTM	Confederación de Trabajadores Mexicanos
	Confederation of Mexican Workers
CUCP	Centro Unificador del Comercio en Pequeño
	Coordinating Center of Small Business
DAAC	Departamento de Asuntos Agrarios y Colonización
	Department of Agrarian Affairs and Colonization
FOR	Federación de Obreros Revolucionarios
	Revolutionary Workers' Federation
FSTSE	Federación de Sindicatos de Trabajadores al Servicio del Estado
	Federation of Government Bureaucrats' Unions
PAN	Partido de Acción Nacional
	National Action Party

PPS	Partido Popular Socialista
	Popular Socialist Party
PRI	Partido Revolucionario Institucional
	Institutionalized Revolutionary Party
SME	Sindicato Mexicano de Electricistas
	Mexican Electrical Workers' Union
SNTE	Sindicato Nacional de Trabajadores de Enseñanza
	National Union of Workers in Education
STFRM	Sindicato de Trabajadores Ferrocarrilleros de la República Mexicana
	Railroad Workers' Union of the Mexican Republic
STMMRM	Sindicato de Trabajadores Mineros y Metalúrgicos de la República Mexicana
	Mine and Metallurgical Workers' Union of the Mexican Republic
STPRM	Sindicato de Trabajadores Petroleros de la República Mexicana
	Petroleum Workers' Union of the Mexican Republic
STRM	Sindicato de Telefonistas de la República Mexicana
	Telephone Workers' Union of the Mexican Republic
USEM	Unión Social de Empresarios Mexicanos
	Social Union of Mexican Entrepreneurs

Introduction   1

Among Latin American nations, Mexico frequently has been regarded as unique. Unlike most Latin American countries, Mexico has succeeded in establishing a stable, viable political system. Common characteristics of the area's politics that have been absent from the Mexican scene for some time include the repeated military overthrow of an elected regime, ephemeral political parties, and the frequent resort to violence for the attainment of political goals.

Despite relative agreement regarding the uniqueness of the Mexican system within Latin America, consensus has been lacking regarding certain aspects of Mexican politics. The main areas of disagreement involve the role of the "official" party, the Partido Revolucionario Institucional (PRI) within the political system, the relative leeway possessed by the Mexican president in the political process, the degree to which the lower classes are incorporated into the political system, and the expected direction of future political change.

Because the official party has controlled the presidency since 1929 and its nearest competitor, the Partido de Acción Nacional (PAN) has received only a small fraction of the electoral support given to the PRI, Mexico has been regarded as a single-party or dominant-party system. This phenomenon, combined with the fact that all major organized interests, with the exception of the business groups, are members of the party, led observers in the early 1960s to conclude that the PRI was the main aggregator of interests in the political system and played a crucial role in the decision-making process.[1] More recent works, however, have deemphasized the role of the PRI in the decision-making process and have stressed instead the party's subservience to the executive branch of the government. According to this view, rather than transmitting demands from the rank-and-file members upward to the Mexican president, the PRI's principal role is to control its members and to forward presidential decisions to its constituents.[2]

All analysts of the Mexican political system view the president as the principal balancer of conflicting and competing interests. Some, however, see the president as a relatively passive individual engaged in the ever more impossible task of reconciling and conciliating these interests in order to attain a broad consensus.[3] Others portray the president as orchestrating or manipulating interest-group demands in order to avoid conflict and to achieve goals set by him and his advisers rather than by the various Mexican publics.[4]

In the same vein, all students of Mexico agree that the benefits resulting from Mexico's impressive economic growth have not been distributed equitably but instead have served to reinforce already existing status and class differences. Some scholars, however, attribute the relative neglect of the lower classes to their exclusion from the political system,[5] while others argue that it is precisely their inclusion in the system that accounts for their condition.[6]

There is general agreement that socioeconomic changes, such as the growth of the middle class, increasing urbanization, and higher levels of literacy and education, will have implications for the future of Mexican politics. Some writers, however, predict increasing democratization of the Mexican political system in the form of greater responsiveness toward citizens, a decline of corruption, and perhaps the eventual evolution of the single-party system into a two- or multiparty system.[7] A second prediction is prolonged stalemate resulting from the inability of the president to forge a consensus from the numerous and competing demands with which he will be besieged.[8] And a third view is that the Mexican political system will be able to control the potentially disturbing effects of socioeconomic change in order to remain essentially an authoritarian regime.[9]

Although the proponents of such conflicting interpretations do not fit perfectly into two camps, in general those who emphasize the importance of the official party stress the constraints under which the Mexican president operates, regard the system as inclusionary and foresee increasing democratization. Those who view the PRI as dependent and principally an institution for the control rather than the representation of interest groups also deemphasize the constraints on the president, view the poor as excluded from the system, and forecast continued authoritarian rule. In summary, the first group approaches the Mexican political system from a basically democratic perspective, while the second group considers the Mexican regime essentially authoritarian.

The democratic interpretation of the Mexican political system is represented primarily by works that were written during the early 1960s when American political scientists were strongly influenced by "group theory" and models of pluralist democracy, both of which emphasized the aggressive and important role of interest groups and political parties in democratic political processes. The competing authoritarian interpretation of the Mexican political system has assumed increasing prominence since the late 1960s, which marked the beginning of a period of disillusionment and perhaps frustration regarding the possibilities for democratic development in so-called late-industrializing and dependent countries. This growing disillusionment corresponds to and reinforces a "paradigm shift" in the literature on political change toward what has been called the authoritarian, corporate or bureaucratic model. In this book, I will treat the Mexican regime as a type of authoritarian system, specifically, as an inclusionary, essentially nonrepressive authoritarian regime.

The model of an authoritarian regime is premised on the assumption that the government or (more accurately) the executive is the independent variable and the complex of interest groups is the dependent variable.[10] In other words, the authority flow in an authoritarian system is the reverse of what supposedly characterizes democratic polities. Instead of responding to and reflecting demands, pressures, and initiatives that originate at lower levels, the executive in an authoritarian regime shapes and manipulates demands emanating from below while enjoying substantial leeway in the determination of the goals that the regime will pursue.

Several of the characteristics of an authoritarian regime reinforce this structure of authority. Interest groups often are created by the regime in anticipation of grass-roots demands for the formation of such groups. Their leaders are selected or approved by the executive, and they are expected to represent the interests of both their rank-and-file members and the executive, with the latter taking precedence in cases of overt conflict. As a result, interest groups in an authoritarian regime tend to play a reactive rather than an initiating role, responding to initiatives from above instead of creating and reflecting demands from below. This kind of political pluralism has been called limited, corporate, or ordered pluralism.[11]

Political participation in an authoritarian regime is kept at a relatively low level, because the system is basically elitist. The masses do have a role to play, however. It essentially involves demonstrations of support in elections, rallies, and parades for the activities and policies of the elites. Thus political participation does not entail independent activity aimed at affecting the decisions of the elites. It is instead the participation of people who are to varying degrees dependent upon the regime and are mainly desirous of receiving whatever the regime is dispensing in return for their support and deference.

The predominant style of rulership in an authoritarian regime is patrimonial or clientelist. The population is organized into interwoven vertical chains of patron-client relationships. The patrons and clients are of unequal status, with the result that vertical chains are multiclass (that is, they consist of individuals from all classes). The closer one is to the top of a particular vertical structure, the higher is one's class. These multiclass chains are held together through the higher status member's distribution of particularistic rewards such as political patronage, money, privileged information, and the like to the lower status member. Such patrimonial rulership, because of its emphasis on vertical linkages, makes difficult the formation of horizontal or class-based alliances in authoritarian regimes.

The value structure that underlies such regimes is, not surprisingly, authoritarian. It is also nonegalitarian. The basic unit of politics is not the individual but the corporate group. Such groups do not all share the same status. Inequality of corporate status and differential treatment of corporate groups are expected and accepted. Available resources are used to reinforce existing inequalities, thus causing authoritarian regimes to be oriented toward status quo politics. Strong leadership is valued and people are deferential toward authority.

While acknowledging the generic similarities among regimes classified as authoritarian, several authors have begun to discuss variations among them. Sometimes these variations are only implied, but there are several explicit attempts to construct typologies of authoritarian regimes.

Perhaps the first suggested basis for a typology was that of Juan Linz. In his pioneering article, "An Authoritarian Regime: Spain," Linz noted that "degrees of mobilization might be the most useful criteria by which to distinguish subtypes of authoritarian regimes."[12] In a subsequent typology, Linz retained the level of mobilization as a possible way of distinguishing among authoritarian regimes, while arguing that a focus on interest groups and their relation to the central authority would constitute a more fruitful approach. Specifically, Linz suggested that "the type and degree of pluralism tolerated and those [groups] excluded from that pluralism should provide . . . a basic dimension for any typology."[13]

A similar argument was made by Guillermo O'Donnell, who developed a threefold typology of authoritarian regimes based in part on whether the regime "attempts to exclude and deactivate the popular sector."[14] A "bureaucratic-authoritarian" regime attempts such exclusion while a "populist-authoritarian" system is dedicated to integrating the masses. O'Donnell's third type is a "traditional-authoritarian" regime, which is neither exclusionary nor inclusionary since it exists in traditional, static, undifferentiated societies.[15] There are therefore no available mobilized masses either to incorporate or to remove from the political arena.

In O'Donnell's typology the decision to include or to exclude hinges on the socioeconomic nature of the group in question. The group that is usually the focus of attention is the so-called popular sector. Linz, however, notes that exclusion or incorporation need not be based on economic considerations, but may be based on cultural cleavages present in the society.[16] Thus, an authoritarian regime can include or exclude groups because of their religion, language, tribe, ethnic identity, and the like. In Latin America, however, where cultural cleavages are not particularly salient, exclusion would probably be based on class, regional, or personalistic criteria.

Another suggested way of differentiating among authoritarian regimes focuses not on the interest groups but on the elites. At issue is the degree of elite consensus or integration, or, to state it somewhat differently, the degree of elite disintegration or competitiveness. Schmitter, for example, in his study of interest group politics in Brazil, discusses degrees of elite competitiveness and argues that Brazil prior to 1964 was "a semi-competitive authoritarian regime."[17] Rogowski and Wasserspring's distinction between "egotistical" and "reciprocal" corporatism also deals basically with the presence or absence of consensus and cooperation. In egotistical corporatism, "competition among corporations is unbridled, or is governed only by the relative physical strength of the competing groups," while in reciprocal corporatism, competition "is restrained by at least some sense of overriding mutual interest in limiting corporation demands."[18] Finally, Kaufman's typology is in part based on differences in the relative strength of the center in comparison to the interest groups, or, in other words, on the degree of elite cohesion at the top.[19]

The question arises, however, of whether a regime that is characterized by weak elite integration at the center can accurately be regarded as authoritarian. At the very least, one would expect an authoritarian regime to be able to make authoritative decisions. A political system dominated by nonconsensual elites seems unlikely to possess such a capability. Rather, such a system would probably be characterized by chronic stalemate, indecision, and stagnation as a result of the elites' inability to agree upon a course of action. According to Huntington, in a modern or modernizing society, the stalemate and stagnation would ultimately lead to political instability, since the demands and expectations of the mobilized groups would overload and thereby cause the disintegration of the existing political institutions. Huntington has labeled such a society "praetorian."[20]

It is conceivable that a praetorian society may possess some structural characteristics that are also present in stable, institutionalized authoritarian polities. Brazil prior to 1964, for example, although lacking a stable center of authority, had an authoritarian political culture, subservient and reactive interest groups, and a patrimonial style of rulership. In a sense, it was a "potential" authoritarian regime. Many of the basic structural components of an authoritarian regime were present. What was lacking was a stable center of authority or, to state it somewhat differently, elite integration at the top. To argue that such a system is authoritarian, however, is erroneous. As Huntington has stated, such political systems "are unclassifiable in terms of any particular governmental form because their distinguishing characteristic is the fragility and fleetingness of all forms of authority."[21] Although they could become authoritarian regimes, they could just as easily remain praetorian or succumb to a revolution that results in the institutionalization of a nonauthoritarian political system.

Although I have questioned the utility of typologies of authoritarian regimes based on the degree of elite integration, typologies that classify authoritarian regimes in terms of their adherence to exlusionary or inclusionary policies seem more promising. Certainly it would appear to make some difference whether an authoritarian regime included all significant mobilized groups and thereby was able to eschew the use of coercion or whether it excluded important groups and consequently had to rely heavily on force in order to maintain itself.

There are, however, several conceptual problems with O'Donnell's typology based on exclusion or inclusion. The authoritarian regimes that O'Donnell classifies as inclusionary or incorporating include regimes like that of pre-1964 Brazil. Although the Brazilian elites were clearly involved in incorporating new groups, principally the lower classes, into the political system, the regime was basically an unstable praetorian regime rather than an authoritarian one.

Furthermore, in his exclusionary category, O'Donnell includes both Brazil since 1964 and Mexico, despite his acknowledgment that the Brazilian regime depends heavily on coercion in order to maintain itself while the Mexican regime does not. Although the Mexican lower classes are not forcibly excluded from and, in fact, are formally included in the authoritarian coalition, O'Donnel contends that they are in reality excluded from sharing in the material and status benefits that are distributed because of the co-optation of their leaders.[22]

The fact that the lower classes in Mexico do not share equally in the country's riches does not mean that they are excluded from the political system, however. In fact, Mexico has been able to achieve political stability with a minimum of coercion precisely because the lower classes have been incorporated into the political system and accept it as legitimate. O'Donnell therefore diminishes the utility of his typology by mixing structural and moral criteria indiscriminately.

Despite this difficulty with O'Donnell's typology, I will retain his distinction between inclusionary and exclusionary authoritarian regimes, although I will redefine the concepts. Both inclusionary and exclusionary authoritarian regimes must be characterized by elite integration at the center. Thus praetorian systems like those of Brazil and Argentina in the early 1960s are not relevant to the typology. An exclusionary authoritarian regime is one that relies heavily on coercion in order forcibly to exclude one or more mobilized groups from participation in the system. Brazil since 1964 is an example of such an authoritarian regime. An inclusionary authoritarian regime, in contrast, does not forcibly exclude any mobilized groups from participation. The legitimacy of such a regime is greater and as a result, reliance on coercion is unnecessary except in extenuating circumstances. The Mexican regime is an example of this inclusionary, essentially nonrepressive type of political authoritarianism.

The expenditure of time and effort that scholars have devoted to developing and refining the authoritarian model is justified by the belief that different kinds of political systems are characterized by different kinds of political processes. Thus political recruitment, socialization, decision making, and participation in an authoritarian regime, for example, are assumed not to follow the same patterns as the equivalent processes in a democratic political system. Accurately classifying a particular regime as authoritarian therefore makes it more likely that information gathered and observations made will be interpreted in ways that conform more closely to the reality of the system rather than to an inappropriate model.

To date, however, considerably more energy has been spent on developing and refining the authoritarian model than in spelling out the implications of classifying a regime as authoritarian. Specifically, there is now substantial agreement regarding the defining traits of an authoritarian regime, as well as a fairly strong consensus regarding the authoritarian nature of the Mexican political system. However, there has been little effort to conceptualize how political processes in authoritarian regimes in general and in the Mexican regime in particular differ from corresponding processes in other kinds of regimes. This deficiency is partially explained by the relative scarcity of detailed empirical studies of political processes in authoritarian regimes.

This book provides the first detailed, in-depth case study of the decision-making process on the national level in Mexico. Although observations and generalizations derived from the study of a single decision may not prove relevant to all decisions made in Mexico, a case study can generate useful insights and information that can be refined or modified by subsequent research. In addition, although Mexico cannot be considered the prototype of all authoritarian regimes, it shares with them a sufficient number of traits to make analyses like this one relevant to understanding the decision-making process in authoritarian regimes in general.

The decision of President López Mateos to implement the provisions for profit sharing of the 1917 Constitution was chosen as the subject of this case study for several reasons. First, it involved the most important interest groups in the country (the organized industrial and commercial interests and the organized labor movement) and their interaction with the executive branch of the government. Because the business groups (referred to as the "private sector" in Mexico) are formally excluded from the dominant political party and the labor groups are officially members of the party, the decision could clarify the role of the PRI in the decision-making process. Second, the decision was chosen because it was particularly complex. It was redistributive because it proposed to transfer profits from the private sector to the organized labor movement. Consequently, it had the potential for generating substantial opposition and thus could illustrate the mechanisms used by the regime to resolve conflict. Third, in view of the increasing criticism to which the Mexican political system has been subjected because of its role in perpetuating inequalities, the decision could help clarify the political obstacles to a more equitable distribution of economic resources. Fourth, the profit-sharing decision embodied regulatory and distributive aspects and therefore could provide insight into a range of decisions made in the Mexican political system. Fifth, because the workers' right to share in the profits of industry was included in the 1917 Constitution elaborated in the aftermath of the Mexican Revolution, the decision was connected with the bases of legitimacy of the regime and could therefore provide insight into the symbolic aspects of Mexican politics.

The study of the decision-making process in a regime like Mexico's presents difficulties that students of more open political systems do not encounter. There are more constraints on the press and on the media in general, thus devaluing somewhat the validity of the information that does appear. More significant, the amount of data on the decision-making process that is made available to the public is much reduced owing to the greater secrecy surrounding the entire process. I therefore have used newspaper stories only to construct the formal or more public aspects of the profit-sharing decision. Much crucial data concerning the private and less publicized aspects of the profit-sharing decision were gathered during personal interviews with most of the significant participants in the decision.[23] More important, however, was the access I was inadvertently given to two folders stored in the archives of the Confederación de Cámaras de Industria (CONCAMIN). Entitled "Reforms to Article 123," these folders were kept by the so-called gran comisión, the leaders of the major organizations of the private sector who had joined together early in 1962 to plan and implement a united strategy, hoping to ensure that the profit-sharing reforms would be as undetrimental as possible to its interests. The two folders contained the minutes of all the meetings of the leaders of the private sector between January 1962 and December 1963, as well as letters reporting all interviews that its representatives had with the Mexican secretary of labor or other government officials. I supplemented the materials in the folders with published and unpublished communications between the private-sector leaders and their member organizations, all of which are on file at the headquarters of either CONCAMIN or the Confederación Patronal de la República Mexicana (COPARMEX). Although I discovered no folder containing equivalent information about behind-the-scenes meetings of the leaders of the labor movement, the published and unpublished communications between the leadership of the Confederación de Trabajadores Mexicanos (CTM) and its rank-and-file members were extremely useful. Another valuable source of information was the three-volume *Memoria de la Primera Comisión* that President López Mateos ordered published because of his pleasure at the successful resolution of the profit-sharing decision. The volumes include transcripts of the debates of the National Profit-Sharing Commission, as well as the major studies submitted to the commision during its meetings between March and December 1963.

The Political Environment 2

The following description of the Mexican regime as authoritarian is organized around the three defining characteristics of the authoritarian model—the limited autonomy of interest groups, the low level of mobilization and its deferential nature, and the predominance of a patrimonial or clientelist style of rulership. This emphasis on three principal traits is necessarily selective. The general literature on the Mexican political system is quite extensive, and there is no need to repeat what has already been well done by others. However, those who are already conversant with the existing literature will still find much that is familiar in the following pages. This chapter is essentially an effort at synthesis. It does not attempt to present new information but rather to reinterpret basic data to highlight the similarities and differences between Mexico and the authoritarian model and to emphasize those aspects of the Mexican system that are relevant to an understanding of the decision-making process.

The New Legitimacy

Throughout the nineteenth century, there was little to distinguish Mexico's politics from that of her neighbors. The end of Spanish colonialism resulted in the formal adoption of a political system modeled after that of the United States. However, attempts to superimpose a liberal-democratic form of government on a rigidly stratified, agrarian, and illiterate society resulted in chronic instability.

During the last quarter of the nineteenth century, Porfirio Díaz succeeded in adapting the formal political system to Mexican reality by establishing an authoritarian regime. In some ways, Mexico's existing regime resembles the one established by Díaz. Power was highly centralized in the hands of Díaz and a group of aging technocrats known as the *cientfficos*. Major power contenders and their followers were kept in check through a system of patrimonal rulership, popularly known as "pan o palo" (bread or the stick). By means of this system, Díaz manipulated Mexico's various interweaving chains of patron-client relationships, preventing the formation of horizontal alliances that could challenge his hegemony. The interests that were incorporated into the system enjoyed limited autonomy. The system was manageable because of the relatively small number of interest groups and the low level of mobilization of the society in general. Much of the population was excluded from the national political system. Those Mexicans for whom the central system was relevant had little input into the political process. Their main job was to be obedient and to provide support in return for the rewards that Díaz distributed among them.

The Díaz regime collapsed in 1910 as a result of its failure to incorporate the new groups brought into existence by Díaz's successful efforts to industrialize the country.[1] The collapse had been precipitated by a call for "effective suffrage and no reelection." This phrase eventually became associated with the Mexican Revolution, the name that has been given to the chaotic years between the fall of Díaz (1910) and the reestablishment of political order (approximately 1920).

The document that best symbolizes the new consensus that legitimizes the "revolutionary" regime is the Constitution of 1917. The fact that it is still in effect today is partially a tribute to the skill with which the members of the constitutional convention succeeded in synthesizing their somewhat contradictory and conflicting values, beliefs, and aspirations.

One set of principles embodied in the Constitution can be labeled liberal-democratic and includes provisions for a federal system of government; the separation of powers among the executive, legislative, and judicial branches; periodic elections; limitations on tenure of office; and a prohibition against the reelection of the president and the consecutive reelection of members of the Chamber of Deputies. These principles, with the exception of limited tenure and the prohibition of the reelection of the president, were introduced into Mexico at the beginning of the nineteenth century and became dominant in the middle of that century.

Although the 1917 Constitution embodied the principle of the separation of powers characteristic of other liberal constitutions, it did not provide for a balance of powers. Instead, it created a strong executive and a relatively weak legislature. This division of power was a reaction to both the chaos that reigned in Mexico following the fall of Díaz, and the failure of the governments of the mid-nineteenth century. The latter had been based on the liberal Constitution of 1857, which had provided for a relatively weak executive and a powerful legislature. As a result of the 1917 Constitution, therefore, "only the President can promulgate a law, by signing it and ordering its publication. He can veto legislation, in toto or by item . . . In the . . . event of Congress overriding a presidential veto . . . there is no constitutional way in which [the President] can be forced to promulgate legislation still repugnant to him."[2] Executive-sponsored legislation submitted to Congress takes precedence over other business. By virtue of the Constitution's implied powers, "the detailed regulations and codes to carry out constitutional mandates or to give administrative effect to general Congressional statutes" are drafted by the executive and are legally binding.[3] As Scott has pointed out, because most Latin American countries, including Mexico, follow the civil law tradition, "the extent of administrative rule-making is much greater than in a country with common law."[4]

Liberal-democratic principles, with their emphasis on a strong executive, are not, however, the only ones that found a place in the 1917 Constitution. Traditional corporate principles are also present. In the course of the Mexican Revolution, new groups with

14

new demands became mobilized, and since these groups were represented at the convention that drafted the Constitution of 1917, their demands were incorporated into the document. There is therefore a special "labor article" (Article 123), which "provides a set of Utopian norms for the conditions and remuneration of Mexican labor," "enjoin[s] on the state the fostering of a strong Mexican labor movement and [gives] the state powers to regulate it." There is also an "agrarian" article (Article 27), which "defines property in terms of 'social function' rather than in accordance with common law precedents" and provides for both small private landholdings and collectivized communal lands.[5] As Frank Tannenbaum has noted:

> By implication, the Constitution recognizes that contemporary Mexican society is divided into classes, and that it is the function of the State to protect one class against another. The Constitution is therefore not merely a body of rules equally applicable to all citizens, but also a body of rules especially designed to benefit and protect given groups. . . . The pattern of the older Spanish State, divided into clergy, nobility and commons, has been recreated in modern dress, with peasants, workers and capitalists replacing the ancient model. This is not done formally but it is done sufficiently well to make it evident that a very different kind of social structure is envisioned in the law, even if only by implicit commitment, than that in a liberal democracy.[6]

The combination of a strong executive and an implicitly corporate organization of society that was sanctioned by the 1917 Constitution was reminiscent not only of the Spanish colonial system of government that had existed in Mexico from the mid-sixteenth century until 1810,[7] but also of the Díaz regime of the late nineteenth century.

The fact that the 1917 Constitution provides for a system of government that resembles past governments is important in Mexico because many Mexicans are still wedded to traditional modes of living and thinking. Such individuals do not obey a government because it is based on laws equally valid for all. Instead they submit to authority because they have always done so. Their obedience is often unconditional and traditional. The outward similarities between the present Mexican regime and past regimes (e.g., the Spanish colonial regime and the Díaz regime)

15

therefore imbue the current regime with traditional as well as legal legitimacy, although traditional legitimacy is mainly important with regard to the more parochial Mexicans.

The Ideological Legacy of the Revolution

The Mexican Revolution produced more than a new legitimizing formula upon which to base a new political system. It also succeeded in shifting the locus of power from the landed elites, who were deprived of their lands in the course of the Revolution, to elements of the new middle class. The Revolution that brought the new middle class to power also provided it with a common identity and a shared ideological perspective regarding the direction of future change and the goals to be pursued. This statement is not meant to imply that the Revolution was an organized, ideologically coherent movement bent on achieving specific, well-defined ends. There was, in fact, no general agreement regarding the "goals" of the Revolution. As Scott has noted, the Revolution meant and means "everything to everybody and something different to each."[8]

It is possible, nevertheless, to isolate several predominant ideological themes that are related to the Revolution and that have come to constitute a kind of ideological elite consensus. One is nationalism. Antiforeign sentiment was an important element of the Mexican Revolution. Under the Díaz dictatorship, foreign penetration and influence, particularly that of Great Britain and the United States, reached such heights that it was often said that Mexico was "the mother of foreigners and the stepmother of Mexicans." The diplomatic intervention of the United States during the early years of the Revolution reinforced such sentiment.

Hatred of the foreigner is not, however, synonymous with nationalism. It was only after the fighting ended and the revolutionary leaders began to consolidate and institutionalize their control that a conscious effort was made to transform the antiforeign sentiment of the Revolution into pronational sentiment. The nationalism that the regime stresses is an eclectic mixture of traditional and modern elements and thus enables all sectors of the population to identify with it. Pride in Mexico's Indian heritage, in her "mixed race," which resulted from the joining of the Spaniard and the Indian, and in her agrarian past is mixed with a future-oriented faith in progress, change, and economic development. The new nationalism is also diffuse enough to enable the regime to justify almost any decision or policy in terms of it.

Another of the Revolution's themes is "constitutionalism," the sanctification of the Constitution of 1917.[9] The Constitution speaks of the separation of powers, religious freedom, liberty, no reelection, and free public education, and such provisions have given impetus to aspirations for democracy among many Mexicans. As a result of these liberal-democratic provisions, "opposition" groups, including minor political parties that pose no threat to the regime, are tolerated and often deliberately sustained by the regime, as long as their criticism is constructive rather than destructive. Constitutionalism also involves a delegation to the government of the power to defend and advance the rights of laborers and peasants, a stipulation that conflicts with the spirit of the liberal-democratic provisions. As a result of these contradictions, "a president can find constitutional justification for just about anything he wishes to do."[10] Constitutionalism, like nationalism, therefore provides the regime with considerable flexibility.

Two other major themes are *justicia social* (social justice) and economic development. Aspects of justicia social include redistribution of income and agricultural lands in favor of the masses, the expansion of free public services, the enforcement of price and rent controls, and the general improvement of the standard of living of the population as a whole. Economic development refers basically to industrial growth. All forms of ownership, whether private, public, or communal, are acceptable, since all can be sanctioned in terms of nationalism, constitutionalism, social justice, or economic development.

Authoritarian Aspects of the Mexican Political System

The Constitution of 1917, embodying both liberal-democratic and corporate principles, and the Revolution's elite ideology are the specifically Mexican underpinnings of a type of political system that is not unique to Mexico. In fact, Mexico shares all the basic structural characteristics of the authoritarian or corporate model outlined in the preceding chapter. The autonomy of her interest groups is limited by the central government; the level of mobilization is basically low and deferential in nature; and patrimonial rulership prevails.

The Limited Autonomy of Interest Groups

The most visible groups in Mexico are organized around agrarian, labor, middle-sector, and business interests. Although the profit-sharing decision involved only the labor and business interests, the peasants and the middle-sector groups will be briefly described in order to clarify the relative position of organized labor and business in the political system as a whole.

The peasant, worker, and middle-sector groups are formal members of the respective agrarian, labor, and popular sectors of the trisector PRI, the "official" party of the regime. The business interests, which are formally excluded from the revolutionary party because of their nonrevolutionary character, are instead obligated to join either CONCAMIN or the Confederación de Cámaras Nacionales de Comercio (CONCANACO).

Although none of these groups is completely autonomous, the strength of their ties to and dependence upon the regime varies. Differences in origins, financial resources, level of education, and degree of political consciousness are some of the factors accounting for variations in the ease with which the executive can manipulate them.

The Peasants. The group that is most easily manipulated are the 2.5 million *ejidatarios*, poor and often illiterate peasants who have been given communal lands (*ejidos*) by the regime. The ejidatarios are dependent upon the regime for the continued possession of their land, for necessary credit and technical assistance, and for protection of their rights in general. All ejidatarios are subject to a detailed Agrarian Code, whose often broad provisions are interpreted and applied by the regime. All are also, with minor exceptions, nominal members of the regime-created and supported Confederación Nacional Campesina (CNC), the principal organization within the agrarian sector of the official party. Although the CNC theoretically is controlled by its membership and represents the interests of the ejidatarios, in reality it is the regime that determines who the CNC's leaders will be and what goals the organization will pursue. This fact does not mean that the CNC's leaders never represent the interests of the peasants. Rather, it signifies that when the needs of the peasants and those of the regime conflict, the regime's interests take precedence.

It is not difficult for the regime to ensure that its candidates win CNC elections. Slates are prearranged. Elections for the members of the CNC executive committee are presided over by a representative of the Departamento de Asuntos Agrarios y Colonización (DAAC) who, according to the law, must approve them in order that they be considered legitimate. The fact that the elected head of the CNC has always been a member of the middle or upper class, usually lacking a rural background, attests to the efficacy of such procedures. On the state and local levels, the elected officials of the CNC frequently are selected by the state governors.[11]

Despite these controls, rebel leaders intent upon mobilizing the *campesinos* in order to improve their socioeconomic and political status have occasionally emerged. In such cases, the regime has either persuaded the dissident leaders to work within the system or, if that effort has failed, incarcerated them. The history of the Central Campesina Independiente (CCI) illustrates both methods of dealing with potential threats to the regime's control of the peasants. Formed in 1963 in northern Mexico by radical members of the CNC, the movement began to decline during the following year when two of its leaders, Alfonso Garzón and Jacinto López, decided to cooperate with the regime and the most important recalcitrant leader, Ramón Danzós Palomino, was jailed for an extended period of time.[12] Peasant leaders whom the regime considers to be extremely threatening have met with more severity. The most notorious case is the still unsolved assassination of peasant leader Rubén Jaramillo and his family in 1962.

The regime's control over CNC policy is derived from its control of the organization's leaders. The CNC statutes provide for a high degree of centralization in the hands of these leaders. State and local elections and meetings to discuss political and social matters cannot be held without authorization by the CNC's national spokesmen. Results of unauthorized activities are considered to be null.[13]

The effectiveness of the regime's control over the peasants is attested to by several facts. Despite their relatively great numbers and high level of organization, the ejidatarios have prospered least of all major organized groups in Mexico. Their lands are often of the poorest quality and lack water. Major irrigation projects continue to benefit private landowners rather than the ejidatarios. Agricultural credit has not been channeled to those who need it most but rather to the relatively small number of prosperous ejidatarios because the latter are regarded as better credit risks. Nor has the government acted to break up large private landholdings in order to redistribute acreage among the growing number of land-hungry peasants.[14]

Organized Labor. Organized labor is not as dependent upon the regime as the organized peasants. Its members are more mobile, more educated, and in theory are not dependent upon the regime in order to exercise their skills.

Approximately one-fifth of the nonagricultural labor force is unionized. This fraction represents a slight decline since 1940 as a result of insufficient effort to organize the rapidly expanding service sector of the economy.[15] (Agricultural workers are almost completely unorganized.) Although estimates vary greatly, all agree that the largest labor organization is the Confederación de Trabajadores Mexicanos (CTM), which was founded in 1936. Next in importance is the Confederación Revolucionaria de Obreros y Campesinos (CROC), which was established in 1952. The Confederación Regional de Obreros Mexicanos (CROM), founded in 1918, is Mexico's oldest surviving labor confederation. The other major organizations are not organized by state or region as are the CTM, CROC, and CROM but by industry. The largest of these national unions are the STFRM (railroad workers), STMMRM (miners and metallurgy workers), and the STPRM (petroleum workers). In the 1950s and early 1960s, all of the labor organizations were divided into two blocs, the CTM-dominated Bloque de Unidad Obrera (BUO) and the CROC-dominated Central Nacional de Trabajadores (CNT). By 1966 the government had succeeded in uniting the labor movement under one umbrella organization called the Congreso del Trabajo.[16] All members of the Congreso also belong to the labor sector of the PRI. (See Table 1.)

Although the organized workers have relatively greater education and mobility than the peasants, however, their independence is seriously circumscribed. All unionized workers are subject to the provisions of the Federal Labor Law (which are interpreted and applied by the regime) and hence are subject to varying degrees of control by the regime. In order for a union to acquire property, negotiate contracts, and represent its members, for example, it must register with the Ministry of Labor. If the regime decides that the union does not meet the legal requirements, it will be denied registration.[17] An unregistered union "loses the protection of the labor law and is completely on its own in collective bargaining and strikes which means that the group probably ceases to function as a trade union."[18] This fate befell the General Union of Mexican Workers and Peasants. Founded in the late 1940s by Vicente Lombardo Toledano, the government refused to grant it recognition. As a result, it eventually lost adherents from the sugar, oil, and mining industries and failed to develop beyond the status of a weak pressure group.[19]

The regime also controls the right to strike by virtue of its right to classify strikes as legal or illegal, and, if legal, as existent or nonexistent. If a strike is declared illegal (because it involves violence or results in property damage, for example), workers are subject to immediate discharge, fines, imprisonment, or a civil suit.[20] If the strike is declared legal but nonexistent, strikers forfeit

Table 1

Estimated Size of Labor Organizations in Mexico

Former affiliates of the BUO	Higher estimate	Lower estimate
CTM.	1,250,000[a]	213,098[b]
CROM.	200,000[a]	23,098[b]
CGT	18,000[a]	9,405[b]
STFRM		101,263[e]
STMMRM	86,000[g]	74,106[e]
STPRM	28-30,000[c]	24,503[e]
STRM	8-10,000[d]	
Former affiliates of the anti-BUO bloc (CNT)		
CROC	400,000[h] & 300,000[a]	32,393[b]
CRT		2,535[b]
SME		10,000[f]

a. Computed by Miller in Richard Ulric Miller, "The Role of Labor Organizations in a Developing Country: The Case of Mexico" (Ph.D. diss., Cornell University, 1966), pp. 45-46, based on statistics from the U.S. Embassy in Mexico City and *Labor Facts in Mexico*, June 1965.
   b. Secretario del Trabajo y Previsión Social, Departamento de Registro de Asociaciones, cited in Guadalupe Rivera Marín, "Los conflictos de trabajo en México, 1937-1950," *El Trimestre Económico* 22 (April-June 1955): 268.
   c. Union sources, cited in Miller, p. 53.
   d. U.S. Labor attaché, U.S. Embassy in Mexico City, cited in Miller, p. 53.
   e. Secretario del Trabajo y Previsión Social, Directoria General de Agrupaciones Sindicales de la República Mexicana, cited in Rivera Marín, p. 485.
   f. U.S. Bureau of Labor Statistics, *Labor in Mexico*, cited in Miller, p. 53.
   g. Interview with Alfredo Rodríguez of the Miners' Union, 1968, Mexico City.
   h. Interview with Enrique Rangel of the CROC, 1967, Mexico City.

rights to lost salaries, "a loss worth considering in the absence of union strike pay."[21] For a strike to be declared legal *and* existent, it must meet certain vaguely defined conditions, such as having the adjustment of the "balance between labor and capital" as its objective.[22] The regime decides whether or not the conditions have been met.

The financial weakness of the labor movement also reinforces its dependence on the government. Union dues equal approximately 1 percent of a worker's salary. Although this percentage is comparable to that contributed by workers in the United States, because of relatively low salaries in Mexico the average Mexican worker pays less than U.S. $1.00 per year to his union.[23] Frequently, dues are not collected or member unions fail to forward the monies to higher levels of their organization.[24] The labor movement consequently relies heavily on government subsidies for its continued viability. Although official data in this regard are nonexistent, one authority estimated that the CTM receives approximately 500,000 pesos annually in direct subsidies; the CROC, about 300,000.[25] These sums do not include the government's significant donations of land, buildings, and equipment to the labor organizations.[26]

Furthermore, the regime can exert substantial leverage on union leaders. Since union leaders, as well as unions, must be registered with the Ministry of Labor, undesirable leaders can be denied registration. Unless the union's leaders are registered, the legal status of its documents becomes questionable. The results of a 1962 strike against a government automobile company, a strike upon which the government did not look favorably, suggest the implications of this arrangement. Because the strike petitions included names of unregistered union leaders, the strike was declared illegal and the workers were ordered to return to work under threat of dismissal.[27]

The regime can also use its vast resources and its control of political patronage to encourage labor leaders to cooperate. The territorial divisions of the major labor confederations closely parallel those of the PRI and of electoral districts. Deserving heads of local labor federations therefore can sit on district committees during elections and serve on a municipal committee of the PRI. Leaders of regional federations can be given positions in federal electoral districts and on the district committees of the PRI, and leaders of state federations can become state legislators and leaders of the party's regional committees. At the highest level, national labor leaders can be awarded seats in the Chamber of Deputies and the Senate, as well as positions of importance in the highest organs of the PRI. Fidel Velázquez, Jesús Yurén, Blas Chumacero, and Sánchez Madariaga, all of the CTM, have each served multiple terms in the Chamber, the Senate, or both.

Once the national leaders are co-opted, the regime's control of subordinate leaders and of the membership in general is facilitated, since the national leaders have the power to recognize or withhold recognition from the lower-level regional leadership. In the absence of such recognition, the regional union cannot be officially recognized and therefore will be ineligible for government protection.[28] In addition, by means of the closed-shop provisions of the Federal Labor Law, union leaders can deprive uncooperative individuals of union membership (thus effectively barring them from employment).[29] This power is significant in a country like Mexico where unemployment and underemployment are relatively high.

Should the regional unions wish to oppose or remove the national leadership, they must overcome numerous obstacles. As has been noted, the major labor confederations are divided into state, regional and municipal federations or into national industrial unions, such as the national union of sugar workers. The regime and co-opted national labor leadership prefer the first kind of organization (which is the predominant one). Because workers in the same kind of industry are distributed among the various regional federations, horizontal contacts among them are discouraged, thereby decreasing the possibilities for successful revolt from below.[30] The fact that the national leaders are not elected by direct vote but instead by a public show-of-hands by carefully chosen delegates further limits opportunities for removal of incumbent leaders. It is not surprising that there is little circulation of elites within the organized labor movement. One man, Fidel Velázquez, has led the largest labor confederation, the CTM, during all but ten years of its existence.

If dissident leaders gain control of their organizations despite the many obstacles, however, the resources of the regime can be used to persuade the new leaders to adopt a more cooperative stance. Should such persuasion fail, intransigent leaders and their

23

followers can be removed from their positions. During the presidency of Alemán, for example, repeated purges of left-wing leaders and imposition of more cooperative individuals occurred.[31] In a more recent example, López Mateos brought the railroad union into line in 1959 by declaring its strike illegal, jailing its leader, Demetrio Vallejo, and purging Vallejo's supporters from the union.[32]

There are numerous indications that the mechanisms for control of the organized labor movement have proved effective. One is the extent of disunity within the labor movement. Despite the absence of significant ideological cleavages among the major groups, cooperation among them is minimal. Scott, for example, notes that the CTM tried to prevent a large turnout at a CROC-sponsored mass meeting in support of the candidacy of López Mateos and also that the secretary general of the CGT asked members of his union to vote for López Mateos for president but not for the CTM congressional candidates.[33] Labor-management conflicts also elicit little cooperation. In the city of Ensenada, for example, Ugalde found no evidence that a labor federation "[had] ever joined another in solidarity strikes or protests even in cases in which the labor groups seemed to have a common cause."[34]

Additional evidence of the extent of the regime's leverage is the strong relation between a president's attitude toward particular labor groups and their growth or decline. The regime's resources and controls frequently have been employed to weaken labor organizations that have become too independent of the president and to establish or strengthen more loyal organizations that can serve as countervailing forces. The CROM, for example, grew rapidly as a result of the support of Presidents Obregón (1920-1924) and Calles (1924-1928) and precipitously declined when President Portes Gil (1928-1930), with the blessing of Calles, withdrew government support from the confederation.[35] Cárdenas (1934-1940) subsequently aided Lombardo Toledano in forming the CTM in order to undercut the strength of labor groups that were loyal to Calles.[36] Approximately one decade later, President Alemán (1946-1952) replaced the left-wing Lombardo with Fidel Velázquez, a man whose views were more akin to his own. Alemán's successor, President Ruiz Cortines (1952-1958), supported the creation of the CROC and "seemed to be playing [it] off against the CTM, . . . perhaps because certain CTM leaders and others in the BUO tended to lean toward [the alemanista] faction in the PRI that sought to restrict Ruiz Cortines."[37] There are also strong indications that López Mateos (1958-1964) fomented the creation of the CNT in the early 1960s to offset the influence of the CTM-dominated BUO.[38] The formation of the Congreso del Trabajo, an organization that includes all labor organizations, during the Díaz Ordaz presidency (1964-1970) was an attempt to strengthen the unity of the labor movement in order to use it as a counterweight to the relatively stronger and more unified private sector.

The number of strikes provides another indication of the dependency of the organized labor movement. As González Casanova has noted, there is little correlation in Mexico between strikes and the economic cycle, with one or two exceptions. However, the number of strikes does correlate with the Mexican president's attitude toward organized labor. In general, the more favorably inclined a president is known to be toward organized labor, the greater the number of strikes.[39] For example, there were approximately 478 strikes per year during the Cárdenas years and 472 strikes per year under López Mateos. Both presidents were regarded as highly sympathetic toward organized labor. During the presidency of Alemán, who "proceeded upon the principle . . . that what was good for Mexican business was probably good for Mexico,"[40] the number of strikes per year equalled 108.[41] One way Alemán kept the number of strikes low was to declare proposed strikes nonexistent, thereby effectively forcing workers to choose between returning to work and suffering severe penalties.[42]

Whether or not a strike is settled in favor of the workers also is generally positively correlated with a president's favorable attitude toward organized labor. Under Cárdenas, for example, 79.8 percent of the strikes that were clearly decided in favor of either labor or business were settled in favor of labor. Under Alemán, the percentage declined to 35.7 percent.[43]

The government's control of the labor movement is reflected in the wage structure. Mexico is atypical of developing nations in this respect: its workers' earnings have experienced rather modest changes over the past few decades. Between 1945 and 1960 (roughly during the presidencies of Alemán and Ruiz Cortines), real wages remained almost constant. In the early 1960s (the years of the López Mateos presidency), wages increased, but "fragmentary evidence indicates that the rate of change may have declined in the most recent years."[44] Thus, the political weakness of labor vis-à-vis the government is also reflected by a wage structure that fluctuates more in accord with political factors than with economic ones. *The Middle-Sector Groups.* It is difficult to generalize about the regime's ability to control and manipulate the numerous middle-sector groups.[45] There are more than fifty such groups. All belong to the Confederación Nacional de Organizaciones Populares (CNOP) the umbrella organization that is synonymous with the "popular" sector of the PRI. The largest, which are listed in Table 2, claim several hundred thousand members each, while the smallest list only several hundred affiliates.[46]

Table 2

The Largest Groups within the CNOP

Name	Claimed membership
FSTSE	680,000
SNTE (also members of the FSTSE)	250,000
CNPP	1,800,000
CNCRM	275,000
CUCP	250,000

SOURCE: David Schers, "The Popular Sector of the Mexican PRI" (Ph.D. diss., University of New Mexico, 1972), pp. 203-204.

The CNOP's heterogeneity also makes generalizations difficult. As a nominally middle-sector organization, it includes members of the liberal professions, white-collar government employees, small businessmen and merchants, craftsmen, students, and the like. However, it also counts among its members the lower-class inhabitants of urban *barrios*, (organized into Federaciones de Colonias Proletarias), poor, small farmers affiliated with the Confederación Nacional de Pequeños Proprietarios Agrícolas, Ganaderos y Forestales (CNPP) and street vendors, sellers of lottery tickets, and similarly employed or underemployed individuals.

The size and heterogeneity of the CNOP, however, contributes to its weakness vis-à-vis the government. It is difficult to unify so large and diversified an organization. As a result, the government is able to play the various groups off against each other, thereby maintaining its control. This "divide and rule" strategy, in fact, accounts for the decision to place government workers, who are members of the Federación de Sindicatos de Trabajadores al Servicio del Estado (FSTSE), into a sector separate from the rest of the labor movement. It also explains why the CNPP, which basically represents agricultural interests, is included with the middle-sector groups rather than with the agrarian sector of the party.

Despite such considerations, it would appear that the larger organizations within the CNOP could conceivably create difficulties for the regime. Several factors, however, militate against this eventuality. First, the largest group within the CNOP, the CNPP, is largely a paper organization. Schers estimates that, despite the leadership's claim of nearly 2 million members, only about 12,000 of its members, "the more prosperous ones," are well organized.[47] The remainder are nominal affiliates who cannot be readily mobilized in support of the group's leaders. Second, the behavior of the 650,000-member FSTSE is circumscribed by the fact that its members, all of whom are government employees, are, in the absence of a merit system, dependent on the regime for their continued employment. The teachers, who are organized into the Sindicato Nacional de Trabajadores de Enseñanza (SNTE) are in a somewhat analagous position since most primary and secondary schools, as well as the major universities, are financially supported by the national government and are subject to varied degrees of federal control.

The members of the smaller organizations within the CNOP are also subject to government control, but for a different reason. As Schers has stated, "Street vendors, shoeshine boys, musicians, all need licensing, and the organizations have control over the licenses."[48] Lack of cooperation with the authorities, therefore, can result in prolonged unemployment.

The "cooperation" that the government seeks from the affiliates of the CNOP in part involves a willingness to participate in demonstrations of popular support for the regime. It is difficult, however, to get lawyers, doctors, and other members of the middle sector to take part in parades and rallies. Their reluctance is another reason why the government has included lower-class associations within the supposedly middle-sector CNOP. The urban poor are both more available and more eager to express support in return for social help. The regime, in turn, can point with pride to the fact that it is supported by *all* social classes, as evidenced by the presence of supposedly middle-sector CNOP groups, as well as agrarian and labor organizations, at public events.

The Business Interests. The business interests (or private sector) are undoubtedly the most independent and least subject to the regime's formal control of all the groups so far discussed. Although the revolutionary rhetoric of the regime deemphasizes its ties with the business groups, the Mexican government "has relied more heavily on the private sector for growth than almost any major country in Latin America."[49] Since 1940 approximately 70 percent of Mexican domestic investment has originated in the private sector.[50] In addition, the private sector currently produces about 37 percent of aggregate domestic production and employs 20 percent of the Mexican labor force.[51] Unlike the peasants and the organized labor movement, therefore, the business interests have the potential for undermining the stability of the regime through decisions that could prove detrimental to the so far continuous and impressive growth of the economy.

Also in contrast with other groups, the private sector is organized into five financially self-supporting organizations. Membership in two of them, CONCAMIN and CONCANACO, is obligatory for firms capitalized at 5,000 pesos or more. Each chamber of industry or chamber of commerce is required by law to pay at least 15 percent of its income to support the parent organization. The size of a chamber's financial contribution determines its voting strength in the national organization. Membership in the remaining threee organizations, COPARMEX, the Asociación de Banqueros Mexicanos (ABM) and the Asociación Mexicana de las Instituciones de Seguridad (AMIS), is voluntary.[52]

Until the mid-1950s, there were important differences of opinion and interest both within and among these peak associations that weakened their ability to influence government policies. CONCANACO favored free trade while CONCAMIN desired government protection *and* nonintervention in industrial production. Within both organizations, there were cleavages based on size of firm, regionalism, and nationalism. COPARMEX, which is technically an employers' union, initially opposed all government policies that benefited the labor movement at the expense of employers. Since the mid-1950s, however, these differences have essentially disappeared, and the private sector has learned to live with the large role the government plays in the economy.[53] The business interests also no longer find government interference in the labor movement objectionable because this policy has produced a relatively docile and malleable labor movement. The recent unity of the private sector is therefore another factor contributing to the relatively greater autonomy of the business interests vis-à-vis the government.

The leaders of the private-sector associations receive no pay for their time-consuming services. Competition for the positions, however, is keen, and turnover rates are high, especially within CONCAMIN and CONCANACO.[54] According to my tally, for example, between 1958 and 1968, forty-four men occupied a total of ninety-four possible one-year leadership positions in CON-CAMIN. Both the competition and the elite turnover attest to the importance of these organizations and their relative freedom from government interference in their internal affairs.

Despite these facts, however, the independence of the private sector is substantially circumscribed by the nature of Mexico's economic system, which can be characterized as a form of state capitalism. As a result of a variety of factors, including the inheritance of concepts of commercial and mercantile law from colonial Spain (a precapitalist and Catholic country) and a relatively late start in beginning the industrialization process, the government has always played an important role in the economy. Consequently, the private sector is not as "private" or free from government influence and control as traditional models of capitalist economies would lead one to expect.

The government, for example, owns or controls the most important industries in the country, including the railroad, telegraph, telephone, electric power, steel, aviation, petroleum, natural gas, and petrochemical industries.[55] A tally of the ownership of the thirty largest firms in Mexico in the mid-1960s produced the results shown in Table 3.[56]

Table 3

Percentage owned by	Top 10 firms	Top 20	Top 30
Mexican government	100.00	88.5	82.2
Mexican private	——	8.7	13.9
Foreign private	——	2.8	3.9

Since state ownership and control is sanctioned and encouraged by nationalist sentiments of both the regime and the population in general, the private sector can do little to challenge effectively the continuous expansion of such ownership and control without appearing to be "unpatriotic" and "unrevolutionary." At the present time, there are an estimated 400 public or mixed public-private enterprises in the country, most of which have been created since 1940.[57] Needless to say, the government's pricing policies in these industries can seriously constrict the options of the private sector. Its decisions regarding the purchase of goods and services for these enterprises also affect the private sector.[58]

The government also plays a significant role in the areas of credit and finance. It is involved in approximately thirty public credit institutions including the Central Bank and Nacional Financiera, the government development bank.[59] Its role is further magnified by the unwillingness of private and public foreign capital, particularly agencies such as the World Bank and the Export-Import Bank, to lend to private business in Mexico.[60] The government can and has used its control of credit "to divert funds from construction, commerce, and real estate speculation toward industry."[61] It also can use the Central Bank to manipulate reserve requirements in order to influence the private sector's rate of investment.[62] The government's taxation powers also can be used to this effect.

The private sector is also greatly affected by the government's policies concerning industrialization. The government sets tariff rates, grants tariff exemptions, and allocates rights to import foreign-made light and heavy goods. Furthermore, despite the recent decrease in direct government investment, investments of government enterprises as a share of gross investment have increased since 1956.[63] The government's decisions regarding how and where to invest therefore are of great significance to the private sector. As Vernon has summed up the situation, "The important point is that the private sector operates in a milieu in which the public sector is in a position to make or break any private firm."[64]

The government, however, has not used its potential power to hurt the private sector. Instead, "beginning with the administration of Lázaro Cárdenas, and particularly since the presidency of Miguel Alemán, government and private industry have cooperated to mutual advantage in what might be termed an 'alliance for profits.'"[65] The government has used its control over imports to protect Mexican firms from foreign competition. Approximately 80 percent of Mexico's imports are now subject to licensing, and, as Hansen notes, "the mere capacity to produce domestically has generally been deemed sufficient reason to suspend the importation of competing products."[66] Devaluation of the peso in 1949 and 1954 also provided a form of protection for domestic industry. Tax rates have been kept low, and exemptions have been readily granted to new industries. Private investment has been encouraged through a ceiling on the nominal rate of interest and "a conscious government choice to finance public sector programs through inflation rather than direct taxation."[67] Finally, public investment has been used in a manner that Hansen characterizes as "bottleneck-breaking." Initially, in the 1930s, public investment was directed toward creating such crucial portions of the infrastructure as roads, irrigation projects, and transportation links. Since 1940 it has been directed toward the long-term financing of such import-substitution industries as iron, steel, and oil.[68]

It is conceivable that the congruence of interests and resulting cooperation between the government and the private sector will lessen in the future. In such a situation, the limits to the relatively greater autonomy enjoyed by the private sector would become more evident. At present, however, the restrictions on the behavior of the private sector can be seen most clearly in those specific cases, like the profit-sharing decision described in this book, in which the government and the business groups have opposing views and the government's view prevails.

Political Participation in Mexico

The Mexican regime encourages a special, limited kind of political participation. Individuals are mobilized periodically on a temporary basis, either to vote in regularly scheduled elections or to attend regime-sponsored demonstrations. The purpose of such events is to allow citizens to show their support and to reinforce the regime's legitimacy, which is based in part on the democratic norms expounded in the Constitution of 1917. When not specifically interested in mobilizing support, the regime is content with passive acceptance of its decisions and generally low levels of mobilization.

Deferential Participation. The success of the regime's mobilization efforts is evident from election results. Since 1929 the official party has controlled the presidency, the national congress, all governorships and state legislatures, and most of the 2,300 mayoralities and municipal councils.[69] Its margins of victory have been extremely wide. In the eight presidential elections since 1929, for example, the PRI has received an average of 88 percent of the total vote cast, with a high of 98.2 percent in 1934 and a low of 74.3 percent in 1952. In 1970 it received 86 percent of the total vote.[70] The PRI's nearest competitor, the PAN, has averaged 10 percent of the total vote since 1952, the first year the party ran its own presidential candidate, with a low of 7.8 percent in 1952 and a high of 14.1 percent in 1970.[71]

The PRI's overwhelming majorities are the result of strenuous efforts to "get out the vote" in rural areas or, stated differently, to mobilize temporarily the relatively poor, illiterate, and tractable rural masses. First, although much of the literature on voting behavior would lead one to expect higher levels of mobilization in urban, economically developed areas than in rural, underdeveloped ones, in Mexico the relationship is precisely the opposite. Participation is inversely related to urbanization and economic development.[72] Second, a higher rate of participation is positively correlated with a higher vote for the PRI.[73] It is not surprising that in the poorest, most rural states, those with the largest number of illiterates, the turnout is highest and the opposition vote rarely exceeds 2 percent of the total vote. In the 1964 presidential election, for example, the non-PRI vote in the states of Chiapas, Guerrero, Hidalgo, Tlaxcala, and Oaxaca, all of which are extremely poor, was 1.1 percent, 3.1 percent, 1.6 percent, and 3.4 percent respectively.[74]

These data are directly relevant to our earlier classification of Mexico as an inclusionary, essentially nonrepressive authoritarian regime. It is difficult to regard as politically excluded large sectors of the population upon whom the elites depend for support. Admittedly these poor, rural inhabitants are excluded from receiving their fair share of the economic benefits that are distributed by the regime. However, their inclusion as deferential political participants is directly related to their economic exclusion. It also accounts for the basically noncoercive nature of Mexican political authoritarianism, since there is no need to repress individuals who are cooperative and demand little of the authorities.

The behavior of congressmen in their capacity as lawmakers parallels that of the general voting population. Between 1935 and

1961, an average of 84 percent of the legislative projects sent to the Chamber of Deputies by the executive received unanimous approval. The relatively few projects sent by the president that were not passed unanimously received majority votes in their favor averaging 97 percent of the votes cast. After 1964, when a constitutional amendment increased the representation of minor parties in the Chamber, the percentage of unanimous votes on executive-sponsored legislation declined to approximately 50 percent by 1967, while some of the executive-sponsored legislation that passed by a majority vote had between sixteen and twenty-seven votes cast against it. This phenomenon was new in Mexico, where in the preceding legislative session the number of negative votes on such legislation never exceeded five.[75]

The compliant behavior of Mexico's lawmakers can be attributed to several factors. First, the PRI's congressmen owe their election to the party leadership's decision to select them as candidates. This dependency, in addition to a constitutional provision that prohibits their immediate reelection, provides them with little incentive to defy the party leadership in order to win support from their supposed constituents. Another factor that reduces their scope for independent action is the legislative calendar. Congress meets for only three months each year, between September and December. The relatively short period available for the consideration of legislation is further decreased by the executive's habit of flooding Congress with most of the major pieces of legislation during the final two weeks of the session. Consequently, extended discussion of proposed legislation is not feasible, and time constraints ensure that no major changes will be made by congressmen.[76]

It is interesting to note that even the *diputados* of the major opposition party, the PAN, are relatively deferential with respect to legislation introduced in the Chamber of Deputies by the executive. Between 1958 and 1961, when the PAN controlled a maximum of 6 of the 162 to 178 seats in the Chamber of Deputies, its congressmen contributed to a unanimous vote on executive-sponsored legislation approximately 70 percent of the time. After the 1964 constitutional amendment increased the representation of minority party deputies in the Chamber, both the PAN and the more radical Partido Popular Socialista (PPS) *diputados* contributed to the unanimous vote on executive-sponsored legislation during the 1964-1967 legislature more than 50 percent of the time.[77]

The negative vote on executive-sponsored legislation rarely involved major pieces of legislation, with the exception of the annual budget, which was voted against by the *panistas* in particular on the grounds that sufficient time to debate the proposed expenditures was never allowed.

Although the Congress is clearly subservient to the executive and basically rubber-stamps legislation sponsored by the executive, it would be incorrect to conclude that the Chamber of Deputies and the Senate play no important role in the Mexican political process. As Padgett has noted, legislative committees do a great deal of work "in gathering information for the executive and in sounding reactions to possible proposals."[78] Legislative proposals by individual congressmen also can provide innovative ideas for future executive-sponsored legislation.[79] In such cases, however, the original idea is always modified somewhat in order to disassociate the presidential initiative from that of the deputy involved, thereby preventing the latter from taking credit for the legislation. The alterations are particularly evident in the case of legislation originally proposed by a member of the PAN or the PPS.

Low Level of Mobilization. The deferential nature of participation is one of the factors that reinforces the hegemony of the political elite. The other is the generally low level of mobilization that characterizes the Mexican political system. It was not always a characteristic of Mexico, however, for in the aftermath of the Revolution, the members of political and parapolitical groups were highly mobilized and politicized. The original statutes of the CTM, for example, opposed all collaboration with the capitalist class and stated that "the final aim of its struggles was the abolition of the capitalist regime."[80] The original statutes of the CNC spoke of the need to "act within a frank spirit of the class struggle—the fruit of the land belongs to those who work it."[81] The 1938 Declaration of Principles of the PRI stated that one of the party's many objectives was "the preparation of the people for the establishment of a workers' democracy as a step toward socialism."[82]

As the regime became more stabilized, the intensity of the mobilization and the degree of politicization decreased as a result of the political elite's conscious efforts to demobilize the population and the people's desire for peace and stability in the aftermath of the violent and chaotic upheaval. Many of the old symbols and slogans, like social justice, agrarian reform, the right to strike, and education for the people, remained; but others, like "the fight against imperialism," "socialist education," and the "revolutionary state," disappeared. They were replaced by such new slogans and symbols as "national unity," "harmony between workers and employers," and "only one road: Mexico."[83] The CTM, after 1947, substituted "for the emancipation of Mexico" for its old slogan "for a classless society,"[84] and the official party eliminated all references to the class struggle and dialectical development and spoke instead of "the fundamental rights of man" and "economic and cultural betterment of peasants, workers and other groups of citizens."[85]

Although the number of people who are members of trade unions, peasant groups, political parties, and interest associations is a matter of some uncertainty, there is general agreement that the reported membership figures are inflated and misleading. Since its membership affiliation campaign of the early 1960s, the PRI reportedly has about 8 million members. If this figure is accurate, 23 percent of the total population[86] (or half the country's adult population)[87] belongs to the PRI. Yet much of this membership is nominal. Trade union members, ejidatarios, government bureaucrats, and teachers, for example, often are registered automatically as party members by their association leader.[88] If they are not automatically registered, they join the PRI because they believe they are under pressure to do so. Few individuals resist incorporation into the official party, since party membership entails few duties other than voting in elections and an occasional appearance at rallies or parades. Refusal to affiliate, on the other hand, has been known to result in reprisals.[89]

The vast majority of the total labor force, more than half of which is rural,[90] is not incorporated into labor unions, ejido groups or the official party. It has been argued that most of these unorganized, unmobilized individuals are "marginal" to or excluded from the Mexican regime.[91] As has already been noted, however, these individuals are very much a part of the current system, since it depends heavily on their docile, unquestioning acceptance and support.

Although mobilization rates today are low most of the time, during elections they can be characterized as moderate. The percentage of the voting age population that votes is a matter of some disagreement, although it ranges between 49 and 57 percent. (See Table 4.) In comparison with other Latin American countries,

Table 4

Voting in Presidential Elections

Presidential election	Percentage of voting-age population voting[a]	Percentage of voting-age population voting[b]
1952	57.8	51.8
1958	49.4	56.2
1964	54.0	52.0
1970	——	56.3

a. Pablo González Casanova, *La democracia en México* (México: Ediciones Era, 1965).
b. Ronald McDonald, *Party Systems and Elections in Latin America* (Chicago: Markham Publishing Co., 1971), p. 252.
SOURCE: I computed these percentages from statistics on "percentage of eligible voters registered" and the "percentage of registered voters voting."

excluding Cuba, Mexico's voting turnout is in the middle range, "with the politically more participant countries clustering around the 70 percent mark [and] the less participant around 33 percent."[92] (See Table 5.) Mexico ranks slightly lower than the United States in voter turnout.[93]

There is a definite trend toward higher levels of electoral mobilization in Mexico. (See Table 6.) In part, the increase results from the addition of women to the national electorate in 1958 and the lowering of the voting age from twenty-one to eighteen in the 1970 presidential election. In the latter election, a record high of 27.6 percent of the total population voted.

The progressively higher levels of electoral mobilization have not threatened the PRI's hegemony, however. First, many of the new voters are rural and are still characterized by deferential attitudes toward the political system. Second, although the increasing PAN vote since 1952 is basically an urban vote, it is heavily concentrated in the Federal District, from which the PAN receives between one-third and one-half of its total national vote, and in a few northern cities like Ensenada. The PRI receives only 12 percent of its total national vote from the Federal District,[94] and, as Ames has noted, between 1952 and 1967 "the more urbanized developed states increased their PRI percentage more than the rural states which were always high-PRI."[95] Thus, one cannot conclude from these data that the PRI must of necessity get weaker as Mexico continues to become more urbanized. The increasing PAN vote

Table 5

Votes Cast as Percentage of Population of Voting Age

Country	Percentage voting	Year
Venezuela	77.7	1968
Bolivia	70.8	1964
Nicaragua	70.7	1963
Argentina	70.1	1963
Dominican Republic	69.8	1966
Costa Rica	67 5	1966
Panama	66.1	1964
Chile	64.5	1964
Uruguay	59 2	1962
MEXICO	52 0	1958
Haiti	48.8	1957
Honduras	43.5	1957
Paraguay	36.3	1958
Brazil	34.2	1960
Peru	33.6	1963
Colombia	33.5	1962
El Salvador	30.8	1962
Ecuador	29.5	1960
Guatemala	18.4	1966

SOURCE: Adapted from Martin C. Needler, *Politics and Society in Mexico* (Albuquerque: University of New Mexico Press, 1971), p. 96.

Table 6

Electoral Turnout in Postrevolutionary Mexico, 1940-1970

Year	Votes cast	Total population	% of total population voting
1940	2,637,582	19,763,000	13.3
1946	2,293,547	22,779,000	10.1
1952	3,651,201	27,415,000	13.3
1958	7,483,403	33,704,000	22.2
1964	9,433,619	41,253,000	22.8
1970	14,027,816	50,718,000	27.6

SOURCE: Adapted from Martin C. Needler, *Politics and Society in Mexico* (Albuquerque: University of New Mexico Press, 1971), p. 97.

could, for example, be overcome by a conscientious effort on the part of the PRI to get more of its urban supporters, particularly those in the Federal District, to the polls. Only 3 percent of the migrants from the countryside who live in Mexican cities with populations exceeding 10,000, for example, prefer opposition parties.[96] The mobilization of urban migrants would appear, therefore, to be a fruitful project for the PRI should it decide to increase its vote in Mexico City and in other areas of PAN strength.

Patrimonial Politics

Mexican society is organized into what Chalmers, in his discussion of Brazilian politics, has called elitist political groupings of vertical linkages based on authority relationships.[97] In Mexico, these groupings are tied together at the national level, and they cut across horizontal structures based on class, interest, and status. Although the vertical hierarchies may sometimes look as if they are based on horizontal linkages (for example, in the case of trade unions or peasant organizations), on closer examination one observes that "the leadership seems to come disproportionately from middle- and upper-class backgrounds" and that these groups "do not reinforce solidarity through an emphasis on exclusivity of membership from particular strata but actively seek to identify with higher status groupings."[98] Thus, as noted earlier, the head of the National Confederation of Peasants has never been a peasant and leaders of the organized labor movement like Fidel Velázquez, Jesús Yurén, Blas Chumacero, and others are full-time bureaucrats and politicians who have not worked as laborers for many decades, if they ever did.

With increasing modernization, the national government has become the principal *patrón* in Mexico, replacing the old-style *caudillos, jefes políticos,* and *caciques,* or transforming them into representatives of the regime in their localities.[99] In urban areas, the regime's representatives are fairly diversified and include bureaucrats, labor leaders, congressmen, and high-ranking PRI officials. Clientelism, *personalismo,* or what George Foster has called "the dyadic contract" between individuals of unequal status[100] therefore are capable of surviving the transition from a rural to an urban, industrial society.

At the apex of the contemporary patrimonial system is the Mexican president, the modern equivalent of the traditional patrimonial leader. When errors are commited, blame usually is attributed to his "incompetent" ministers. When attempts to democratize the PRI proved politically disruptive, for example, Carlos Madrazo, the head of the PRI, received all the blame. Although Madrazo, a presidential appointee whose tenure in office was entirely dependent on the president's will, could never have undertaken such a project without the approval and perhaps the encouragement of the president, the latter received no criticism. Although he remains beyond reproach, the president is regarded as the source of all good things and is praised accordingly.

As the source of all benefits and rewards, the Mexican president is forever deluged with petitions for favors by people in all stations of life. Mexican newspapers are full of photographs and stories reporting the numerous personal audiences that the president has granted to labor, business, agricultural, and party leaders, as well as to delegations of peasants, workers, and other "common people." During such audiences the president is presented with a list of the needs of the "subjects" and their dependents.[101] The arduous political campaign of the PRI's presidential candidate prior to every election provides numerous examples of such petitioning. In every town, village, or city that the candidate visits, he is besieged with lists of requests from the inhabitants.

It is, of course, physically impossible for the Mexican president to grant personal audiences to all his subjects. For this reason, petitioners have had to avail themselves of the services of individuals who serve as intermediaries between the president and people seeking his favors. In Mexico, these intermediaries, or "brokers" are variously called *padrinos, valedores, tatas, compadritos, coyotes, influentes,* and *palancas.*[102] Some of the terms have pejorative connotations. The job of these intermediaries is to petition the president or individuals who have access to him on behalf of the individuals who are dependent upon them. Local studies contain numerous descriptions of attempts to locate and deal with such brokers. The new mayor of Ciudad Juárez, for example, used his family connections and contacts in Mexico City in order to obtain positive action on his many requests for aid.[103] In Atencingo, the ejidatarios devoted substantial effort to locating and using allies high in the federal bureaucracy to press their demands for the right to cultivate crops other than sugar.[104] The two men in charge of getting aid for the construction of a water project for the Indian town of Mitla wrote letters to the agency in Mexico City responsible for handling such requests. They also sent boxes of cheese and pork to the man in charge. To make sure their gifts reached the right man, they sent identical gifts to the official's home.[105]

As Wolf has pointed out, the broker's position is a difficult one, for he must be able to offer his local supporters something in the form of a special relationship with outside authorities and, in addition, he must be able to offer the outside authorities organized political support of a kind they could not get without him.[106] However, since most of the intermediaries are themselves dependent on the regime for the benefits they have promised to deliver to their followers, they feel more accountable to the regime than to their followers. Furthermore, because the masses are often not in a position to hold their intermediaries accountable (but are in a position to be manipulated by them), the intermediaries have many opportunities for personal aggrandizement.[107]

Although the existence of intermediaries or brokers somewhat simplifies the task of ruling for the Mexican president, the number of intermediaries and the complexity of modern Mexican society make it difficult for the president personally to supervise all activities. He therefore is obliged to delegate some of his power to his ministers, the heads of the decentralized agencies, and similarly located individuals, while retaining veto power over their actions. Nevertheless, the ministers, the equivalent of Weber's patrimonial retainers, tend to "appropriate" some of the ruler's powers to themselves.[108] As one student of Mexican politics has noted, "the different ministries tend to convert themselves into personal feudal kingdoms of the respective ministers . . . The ministers surround themselves with their most intimate friends, their relatives and their supporters; thus, they can run their ministries with a minimum of outside intervention."[109]

The president, to retain control, avails himself of a "divide and rule" procedure, playing off the various ministers against each other. In addition, the Mexican constitutional provision against reelection of the president ensures that the appropriation of the president's powers by his "retainers" is only temporary, since after every presidential election (i.e., every six years), most of the ministers are changed and the powers that the previous president delegated to them revert to the presidency.[110] Finally, the possibility of permanent appropriation by subordinates of delegated presidential power is decreased as a result of the contradictory provisions of the 1917 Constitution. These provisions enable the president to delegate decision-making power to subordinates in the knowledge that they ultimately will be obliged to turn to him for a definite resolution of the conflicting directives.[111]

All authoritarian regimes, because of the ruler's need to keep political pluralism limited and thus maintain supreme control, are characterized by the predominance of patrimonial rulership over more formalized and institutionalized ways of governing. It may be, however, that patrimonial rulership in Mexico is more exaggerated than it is in authoritarian regimes of more developed countries. One reason is that, in developing countries like Mexico, the central government exercises a great deal of influence over the economy and, by extension, over the groups that are most affected by the government's economic decisions. Furthermore, the scarcity of resources and capital in Mexico means that there are many competitors for the government's favors. Because it is impossible to satisfy all demands, personal "connections" become an important factor in distributive decisions and patrimonial rulership is reinforced.[112]

The Official Party

The existence of an official political party is not a necessary characteristic of an authoritarian regime. This description of the Mexican political system, however, would be incomplete without some discussion of the dominant party, especially because the PRI did play a role in the profit-sharing decision.

Mexico's official political party was established in 1929, nineteen years after the outbreak of the Mexican Revolution and twelve years after the introduction of "constitutional" government. Its original purpose was the consolidation of the control of Plutarco Elías Calles, a revolutionary "chief," and his supporters who exercised de facto control of various regions in Mexico. Soon after its formation, Calles proceeded to strengthen the institutions of the central government at the expense of the regional powerholders. In 1936, under the regime of Lázaro Cárdenas, new groups of peasants and workers were mobilized, incorporated into the party, and were subsequently demobilized.[113] The party was reorganized along functional lines into four sectors—labor, agrarian, popular (middle-sector), and military (the military sector was disbanded in 1940). Since that time, the supremacy of the central regime over both regional interests and the functional interest groups incorporated into the party has been an accomplished fact.

One way of describing the PRI today is to compare it to a United States political machine, since in many ways it resembles the classical machine model.[114] The ways in which the PRI differs from the model can be attributed to the limited political pluralism, the low level and deferential nature of mobilization, and the patrimonial style of rulership that characterize the environment in which the PRI operates.

The machine-like traits of the PRI are numerous and striking. The PRI, like its classical counterpart, is engaged in weakening or avoiding the formation of horizontal class or interest-based alliances among members of the lower economic strata by dispensing services to individuals in exchange for their votes, a process that is particularly evident in urban areas. Studies of lower-class barrios detail how the party sponsors medical clinics, beauty salons, child-care centers, typing and home economics classes; provides free legal services and cheap heating fuel; helps residents to legalize their property titles; and distributes low-cost food.[115] In return, the recipients of the services attend political rallies and *fiestas*, participate in parades and give their overwhelming support on election day to the PRI's candidates.

Also reminiscent of the United States machines is the pervasive *personalismo* of the PRI, which humanizes the government and thereby strengthens the ties between it and the electorate. Numerous local studies provide evidence of the importance of locating individuals within the party or government bureaucracy in order to obtain favorable action on petitions or requests.[116]

The PRI's use and distribution of patronage bears certain similarities to that of the machines in the United States. Instead of fractionalizing the electorate by playing "ethnic arithmetic," however, the PRI mitigates conflict by playing "functional arithmetic." Patronage is distributed "from the top down" to members of the labor movement, the agrarian movement, and the popular sector in accordance with their relative strength within a given locality.[117] The PRI's seats in Congress are also divided among the labor, agrarian, and popular sectors of the party, with the popular sector, which is the largest, obtaining the greatest number of seats.[118]

A final similarity between the PRI and the classical machine is its use of electoral fraud, corruption, and bribery to maintain control. Perhaps the best-known recent examples are the 1968 municipal elections in Tijuana and Mexicali, which were reputedly won by the PAN. The results were annulled by the government because of "irregularities."[119]

Despite these similarities between the PRI and the classical machine, the ways in which the two differ are significant. One of the outstanding characteristics of the American machine was its strong organization, an attribute the PRI seems to share in theory but not in practice. Of the 31,000 municipal committees that the PRI claims it has, for example, only 25 percent meet regularly.[120] Local studies confirm these figures.[121]

In areas where a PRI organization does exist, the party clearly is subservient to the government bureaucracy, a situation totally different from that of the American political machine. In Jalapa, for example, "local PRI officers respond primarily to government initiatives . . ."[122] On the village level, Mundale noted that the role of the president of the PRI in Cerro Grande was largely limited to "his appearances as a chauffeur and companion for the municipal president," an observation that led him to conclude that "the supremacy of the executive over party power is as obvious in Cerro Grande as it is in Mexico City."[123]

The scope and intensity of the PRI's activity also sharply differentiate it from the traditional United States machine. In fact, the only period of sustained activity seems to coincide with elections, when the PRI makes a concerted effort to "get out the vote." Between elections, the PRI as an organization performs few visible functions. Anything it does is also done by such institutions as labor and peasant organizations or by branches of the government bureaucracy, often with greater efficiency.[124] This fact helps explain why the party can survive with a total staff of fewer than 20,000 individuals.[125]

The nature of the "boss" figure is another striking difference between the PRI and the political machine in the United States. The power of the American boss did not depend on his holding public office. In Mexico, in contrast, "the occupancy of a particular office is a prime requisite for power, and this office is secured through the PRI."[126] In addition, in Mexico the boss figure is institutionalized by constitutional rules that prohibit reelection so that the dependence on a particular individual is absent.

The PRI's relative weakness as an organization is evident even in political recruitment. First, the party shares the recruitment function with other institutions with the result that "those wishing to hold political office do not know the best way to go about obtaining it."[127] Second, there is increasing evidence that these other institutions are playing a larger role than the PRI in the recruitment process. A recent study of the recruitment of Cabinet members since 1935, for example, shows that the number of "políticos" serving in the Cabinet has been decreasing steadily over time, while the number of bureaucrats has been increasing.[128] With regard to gubernatorial candidates, there is a similar trend—an "increased tendency since 1940 to recruit [individuals] with less political and more administrative experience."[129] The fact that few Mexican presidents have been active in the PRI prior to assuming the presidency is still another reflection of the party's limited role in political recruitment. No Mexican president has ever headed a labor or ejidatario confederation. Cárdenas was the only Mexican president who ever headed the party. López Mateos was secretary general of the PRI. However, both men occupied their positions for a short time only.[130] The men who have served as party leader have usually been military men or civilians who were not well known to the public even after their appointment.[131]

The relatively weak organization of the PRI, its subservience to the government bureaucracy, its lack of a boss figure, and its narrow scope and reduced activity can be explained by reference to the specific context in which the PRI functions. In general, this context has served to deemphasize the organizational aspects of the party while emphasizing its role as a framework for distribution of rewards and the conciliation of conflicts.

There are several reasons why the organizational aspects of the PRI are less important than those of the American machines. First, the PRI operates in a context in which a viable opposition party on the national level is lacking. In the absence of a credible challenge to its hegemony, the PRI does not need a strong and active party organization. Furthermore, it is difficult to distinguish an undefeated national party from the government. The blurred boundaries between the PRI and the government reinforce the PRI's organizational weakness, since the PRI becomes the recipient of the votes of blindly loyal Mexicans in rural areas. In such a situation, a strong party organization is not needed. As Foster noted of Tzintzuntzan, an area not noted for its strong PRI organization, "elections here are, above all, a device to permit people to participate in a patriotic manifestation, to declare their loyalty to Mexico and to their community and to express their pride in being Mexican."[132]

The low level and deferential nature of mobilization in Mexico also deemphasize the party's organizational aspects and reinforce its function as a framework for distribution and conciliation. Because most individuals are oriented toward the output side of politics and because the regime is not inclined to mobilize the masses on a continuous basis, there is little need for a strong party organization. Authoritarian values as well as the limited pluralism or lack of autonomy of the various interest organizations within the PRI constitute additional incentives to deemphasize the organizational aspects of the party.[133]

Patrimonial rulership makes its own contribution to the organizational weakness of the PRI. Because of the vertical multiclass and multistatus linkage structures, the PRI does not have to deal with as many individuals as did the American machines. Instead it can work with selected individuals (brokers or intermediaries) who are strategically placed in the linkage structure, with the knowledge that the latter will deliver the support of their followers to the party. The party therefore does not always require a separate organizational structure; it can use personal contacts to work through other organizations and institutions. Beals accurately described this phenomenon when he observed that the members of the CTM in Cherán "functioned as a local branch of the official party" and that "from the standpoint of the union and party officials in the State capital, the town [was] organized; from the standpoint of the majority of the inhabitants of Cherán, it [was] not."[134]

It is mainly in those areas where the vertical linkage stucture is weak or absent that the PRI functions most as an organization. It is not a coincidence that the PRI is strongest as an organization in urban barrios, where the residents are recent migrants to the city who usually are not yet integrated into the vertical linkage structures characteristic of the society at large.

The fact that Mexican society is organized into interwoven vertical hierarchies makes the conciliatory function of the PRI somewhat difficult. It is easier to keep track of a few easily identifiable ethnic blocs than it is to manipulate numerous multiclass and multi-interest vertical structures. The conciliation function of the PRI, therefore, is an extremely time-consuming effort, one that goes on behind the scenes and involves personal rather than organizational capabilities. That the conciliation function is difficult and not always effective is evident from its numerous failures. Sometimes a coveted nomination is given to the wrong person, as in Uruapan, where the dissatisfied inhabitants gave their support to the PAN candidate as a result.[135] Sometimes the inability to satisfy all contenders results in the loss of an election *and* violence, as in the Sonora elections of 1967.[136] To date, however, the instances of successful conciliation far outnumber the failures.

Historical Aspects of the Profit-Sharing Issue   3

On December 27, 1961, President Adolfo López Mateos sent an amendment to Article 123 of the Mexican Constitution to Congress, thus formally initiating the decision-making process that was to result in the establishment of a national profit-sharing system in Mexico. The most important provisions of the proposed amendment to the so-called labor article of the Constitution changed the procedure for the establishment of minimum wages, provided added protection to workers unjustly dismissed by their employers and gave the federal government sole power to establish an obligatory profit-sharing system.[1]

With regard to profit-sharing, the amendment called for the creation of a tripartite national profit-sharing commission to be composed of labor, business, and government representatives. The commission was to elaborate a national obligatory profit-sharing system after it had undertaken investigations of the Mexican economy.

The clauses of Article 123 that were affected by the profit-sharing provisions of the 1961 amendment were Clauses VI and IX. Clause VI originally stated that "in every agricultural, commercial, manufacturing or mining enterprise, the workers shall have the right to a share of the profits, which will be regulated as indicated in Clause IX."[2] Clause IX stipulated that "the determination of the type of minimum wage and profit-sharing referred to in Clause VI will be made by special commissions that will be formed in each municipio, to be subordinated to the Central Junta of Conciliation and Arbitration that will be established in each state."[3] The 1961 amendment, by providing for the establishment of a national profit-sharing commission, proposed to transfer jurisdiction regarding profit sharing from the special municipio commissions mentioned in Clause IX to the federal government.

The removal of jurisdiction over profit sharing from the municipal commissions was understandable because they had rarely functioned. The workers' right to share in the profits of industry had been incorporated into the Constitution as a result of the demands of a number of working-class delegates to the 1916-1917 constitutional convention.[4] The demands were somewhat romantic and idealistic in view of the fact that little industry existed in Mexico at that time. One reason, therefore, for the inactivity of the municipal commissions was the virtual nonexistence of industry and the consequent lack of profits to be shared. Furthermore, prior to 1929, in order for the municipal commissions to function, the respective states first had to promulgate state labor laws incorporating the workers' right to share in profits. Nine states, in addition to the Federal District and federal territories, never promulgated a state labor law.[5] The remaining states promulgated labor laws that incorporated the workers' right to share in the profits of industry.[6] These laws, however, remained unenforced, partly as the result of business opposition. Another significant cause was the absence of decisive, positive action by the national government with regard to profit sharing.

By the late 1920s there was no longer any doubt that the states were unwilling or unable to exercise their constitutional right to legislate in labor matters, including profit sharing. In 1929 the Constitution was amended to give the federal government, which until then had shared its power to enact labor legislation with the state governments, exclusive power in this regard.[7] The government then began drafting a federal labor law that was to provide the specific detailed legislation necessary for the federal government to implement the labor provisions of Article 123 of the Constitution.

The labor law was submitted to Congress in 1931 (and remained in effect until May 1970, when a new labor law was adopted), but the status of the profit-sharing provisions of Article 123 remained confused. On the one hand, Article 123 of the Constitution stipulated that municipal commissions had jurisdiction over profit sharing. On the other hand, the 1929 constitutional amendment had given all jurisdiction over labor matters to the federal government. The Federal Labor Law of 1931, however, which provided the detailed legislation for the implementation of the general principles embodied in Article 123 of the Constitution, made no mention of profit sharing. As a result, the profit-sharing provisions of Article 123 were unenforceable. They remained unenforceable until December 1963, when the National Profit-Sharing Commission, established as a result of President López Mateos's 1961 amendment to Article 123, issued its profit-sharing resolution.

Profit Sharing and Interest-Group Demands

Although the specifics of the profit-sharing provisions of the 1961 amendment can be understood as an attempt to make a previously unenforceable constitutional right enforceable, it is not clear why President López Mateos decided to enforce the profit-sharing provisions at all. If no Mexican president before him had made a serious attempt to implement the profit-sharing provisions, why had López Mateos decided to do so?

According to President López Mateos's "Statement of Motives," which accompanied the proposed legislation to Congress, an important reason for his decision was that profit sharing was "one of the legitimate aspirations of the working class."[8] A brief examination of the attitudes of organized labor, however, leads one to question the strength of its desire for profit sharing.

Disunity within the Labor Movement

The fact that the profit-sharing provisions of the Constitution could not be enforced without additional government action had not caused a great deal of concern among the members of the labor movement. Throughout the 1930s, when the Mexican labor movement was relatively revolutionary, most labor leaders were opposed to profit sharing, not because they were uninterested in having labor receive a share of industry's profits, but because they thought that profit sharing implied collaboration between owners and workers and thus would ultimately weaken the workers' class consciousness and undermine their revolutionary spirit.[9] The leaders also feared that profit sharing would sabotage labor's efforts to obtain greater concessions from employers in collective bargaining situations, since the employers would claim that such concessions were superfluous in view of the workers' right to receive a guaranteed percentage of the firm's profits.[10]

By 1950, however, the situation had changed somewhat. The labor movement had grown less radical, and the profits that employers had begun to reap from Mexico's new industrialization had become more conspicuous. As a result, some labor leaders began to view profit sharing in a more favorable light. The degree of enthusiasm that the various labor leaders expressed toward the idea of profit sharing varied, however. The more radical leaders no longer registered categorical opposition to profit sharing, but neither did they make any effort to have it established. The more moderate leaders began to demand the enforcement of the profit-sharing provisions of the Constitution, yet this demand never ranked very high on their list of priorities. It was ranked even lower by the rank and file, who were more interested in obtaining higher wages and low-cost housing, to cite just two examples.

It might be noted that the incorporation of the workers' right to profit sharing into Article 123 of the revolutionary Constitution put the labor leaders in a difficult situation in the 1930s. It was hard for organized labor to reconcile its opposition to profit sharing with the fact that all rights granted to labor in Article 123 of the 1917 Constitution were theoretically very much in labor's interests. As a result, the more radical labor leaders refrained from publicly speaking against profit sharing and from criticizing those labor leaders who took a more moderate position. By 1950, when the more radical leaders began to view profit sharing more favorably, the fact that profit sharing was a right granted by Article 123 caused them to attempt to appear more enthusiastic toward it than they in fact were. Labor leaders who in the 1930s were not opposed to profit sharing and who by 1950 genuinely favored it also tended to exaggerate their actual desire for it because it was provided for in the "labor article" of the Constitution.

In view of these historical precedents, it is not surprising that until 1950 there was little evidence of labor interest in the profit-sharing issue. The 1920 Pan American Labor Congress held in Mexico City produced a petition sponsored by some CROM members that asked the Mexican legislature to take action to enforce the profit-sharing provisions of the 1917 Constitution. Opinion on the matter was so divided, however, that the petition was not acted upon.[11] At the 1934 Congress of Industrial Law, the Federation of Workers' Syndicates of the Federal District presented a petition that theoretically concerned profit sharing; however, the proposal did not really involve a profit-sharing system, but a workers' health insurance program to be financed by employers.[12] The interest in profit sharing expressed fifteen years later at the 1949 Mexican Congress of Labor and Social Security Law was somewhat greater. Six unions presented petitions dealing with profit sharing. Three of them suggested that the workers receive a specified percentage of an industry's profits, one suggested that each worker be given a month's salary as his share, and the remaining two asked only that the profit-sharing provisions of the Constitution be enforced.[13]

In addition to these isolated petitions in labor congresses, a very small minority of the labor organizations in the country tried to influence Congress to implement the profit-sharing provisions of the Constitution. In 1936, for example, the Sole Union of Workers in Commerce, Industry, Banks, Private Offices, and Similar Establishments sent a letter to the Chamber of Deputies requesting that in the process of reforming the Federal Labor Law, the Chamber should include a special section dealing with profit sharing. In 1938 a campaign to deluge Congress with letters demanding the establishment of profit sharing was organized by the Union of Industrial Workers "Power and Progress" of Baja California. As a result, Congress received letters from about ten labor groups requesting that the profit-sharing provisions of the Constitution be enforced. No action was taken by Congress until 1964, when it decided that the letters were irrelevant since President López Mateos had done what was requested in 1961.[14]

Beginning in 1950, for reasons already noted, an increased number of labor leaders began to speak of the workers' unenforced constitutional right to profit sharing and to demand that the government take the necessary steps in order that the workers receive their fair share of industry's profits. The most concerted and serious effort was made by the leaders of the CTM. The only other labor organization to concern itself with profit sharing was the CROM, but it was involved to a much lesser extent.

The history of the CTM's commitment to profit sharing begins with the formation of the confederation in 1936. Profit sharing was one of the goals specifically mentioned in the CTM constitution. This ostensible commitment, however, did not accurately reflect the sentiments of the CTM leaders who, at that time, "did not support the entrepreneur concept that collaboration among classes should replace the class struggle."[15] Although there were no doubt some CTM leaders who truly favored the establishment of a profit-sharing system, most did not at that time. The reference to the workers' right to share in the profits of industry was less a reflection of true sentiment in its favor than a result of the fact that Article 123 of the Constitution granted the workers the right of profit sharing. The CTM leaders were following the lead of the CROM in this regard, for the CROM had included a reference to profit sharing in its constitution in 1918, despite the fact that many CROM leaders had also at that time been opposed to the principle of profit sharing.[16]

The CTM did nothing with regard to profit sharing until the more leftist leaders were purged from the confederation in 1947. As a result of the change in leadership, the CTM slowly began to deemphasize the class struggle and stress more material and tangible goals for the workers. The worsening economic situation of organized labor throughout the 1950s no doubt contributed to this shift in emphasis. In 1950, at the Thirty-Ninth National Council of the CTM, the confederation resolved for the first time to fight for the enforcement of the profit-sharing provisions of the Constitution. The resolution sought to forestall criticism from the CTM's more radical members by stating that profit sharing would not interfere with the class struggle and should be regarded as "only one of several forms of retribution for services rendered, as a supplement to one's salary . . ." It recommended that the CTM send a request to President Alemán (1946-1952) asking him to send legislation to Congress that would implement the profit-sharing provisions of the Constitution. The resolution also stated that member organizations of the CTM and other Mexican labor groups should pressure Congress to act on the president's anticipated profit-sharing reforms.[17]

The CTM leaders anticipated a favorable response from President Alemán because he had "enthusiastically and sympathetically received the resolution of the National Council of the CTM" and had offered to make a detailed study of the demands.[18] The fact that Alemán had engineered the purge of the left-wing CTM leaders and had supported the current CTM leaders contributed to the leaders' confidence that he would be receptive to their demands. When nothing happened after a lapse of more than a year, however, the CTM leaders altered their strategy and began to encourage their member unions to demand profit sharing in their collective contracts.[19] The profit-sharing issue was also deliberately kept alive between 1950 and 1952 by the CTM leaders, who took care to make some mention of it during each of its National Council meetings (held twice yearly).

A further indication that the CTM had decided not to rely on President Alemán for the implementation of the profit-sharing reforms was the fact that in September 1951 the CTM presented its own reform project to Congress.[20] The proposed legislation provided for the suppression of the municipal commissions that were empowered by the Constitution to establish the local profit-sharing systems and stipulated instead that the workers' right to share in profits should be incorporated into collective contracts. If the contracts contained no profit-sharing provisons, workers would be entitled to a minimum of 10 percent of a firm's profits, based on the employer's income-tax declaration to the Treasury.[21]

The CTM reform proposals were given to the relevant committees in the Chamber of Deputies. These committees decided to focus upon the general question of whether or not to establish an obligatory profit-sharing system, rather than upon the specific merits or defects of the profit-sharing system proposed by the CTM leaders, and invited representatives of the principal labor and business organizations to express their opinions. In December 1953 the committees reported that their studies had not been completed and would continue into the next session of Congress.[22]

Between the time the reforms were submitted in 1951 and the 1953 committee report, the CTM leaders constantly berated the congressional committees for their slow action on the reforms[23] and spoke of the broad support that many groups in society had expressed for the reforms. Among its supporters the CTM listed "many deputies,"[24] "almost all the other workers in the country," labor lawyers and economists, "many famous lawyers who represent the business sector," and "prominent members of the Mexican clergy, who, in spite of having different reasons for their support, are in agreement with regard to the fundamentals."[25] The only opponents of the CTM reforms, according to the CTM leaders, were "those who have always opposed the interests of the working class."[26]

The CTM leaders' claim that the labor movement was united behind their profit-sharing efforts were both misleading and exaggerated. Although, for reasons already mentioned, none of the leaders of the other principal labor organizations vociferously opposed the idea of profit sharing, most were less than enthusiastic about it. These leaders publicly expressed doubt regarding the wisdom of establishing the profit-sharing system proposed by the CTM. The Miners' Union, the Electricians' Union, and the Confederación General de Trabajadores (CGT), among others, were not in favor of the CTM proposal.[27] The CROM, when called before the congressional committee to present its views, affirmed its support of the principle of profit sharing, but feared that the CTM project would not provide sufficient protection for the workers. The CROM suggested that the tax declarations on which profit sharing was to be based should be made available to the workers.[28] The CROM considered the CTM its main rival within the labor movement and blamed its own loss of influence on the CTM leaders who defected from the CROM in 1936 to form the CTM. Thus its criticism of the CTM proposal was not surprising. The CROM's criticism was also related to the fact that it had prepared its own profit-sharing proposal, which differed substantially from that of the CTM. The essence of the CROM system was that the employers were to use a percentage of their profits to cover the cost of a social security system for workers. The CROM proposals were never sent to Congress. They were presented instead to President Alemán, who did not act upon them.[29] The more radical CROC, another rival of the CTM (and of the CROM) also opposed the CTM project, specifically because it included profit sharing as part of a collective contract. The CROC felt that such a procedure would "weaken the tactics of union struggle and create in the soul of the worker a sentiment of appreciation for the employer."[30] In addition, because of the lack of sufficient "maturity" among Mexican workers and the distressing conduct of some leaders (no doubt a reference to those of the

CTM), the CROC contended that profit sharing could become a source of increasing corruption. Finally, the CROC stated that a profit-sharing system that did not give the workers the right to intervene in the administration of a business in order to know its financial condition was illusory.[31]

The congressional committees studying the CTM profit-sharing proposal did not produce the favorable report demanded and awaited by the CTM leaders. They claimed that additional study was required. Nothing was heard about the CTM proposal for several years. In the meantime, the CTM leaders attempted to keep the profit-sharing issue alive by alluding to their proposed reforms to Article 123, which included the implementation of the profit-sharing provisions of the Constitution. Throughout the early 1950s, at almost every National Council meeting of the CTM, the CTM leadership spoke of the need to establish profit sharing and asked Congress to approve the CTM reforms.[32] The CTM leaders' periodic meetings with President Ruiz Cortines (1952-1958) were used as a forum to express their desire to have profit sharing established. After such meetings, the CTM leaders tended to speak of the president's "favorable point of view,"[33] yet no presidential effort to establish profit sharing was ever forthcoming during the Ruiz Cortines administration. During the administration of President Alemán, the CTM leaders had also claimed that the president had looked favorably upon their profit-sharing demands and Alemán had also failed to implement the profit-sharing clauses of the Constitution. One may therefore conclude that neither president had any real intention of establishing a profit-sharing system, and their professions of sympathy for the CTM efforts constituted no more than courteous and politically expedient responses.

In October 1956, five years after the CTM had first introduced its reforms in Congress and three years after the relevant congressional committees had announced that additional study of them would be required, the CTM deputies criticized the committees for their failure to report on the proposed legislation.[34] The following year a new series of congressional investigations regarding the attitudes of labor and business toward profit sharing and the other proposals began. It is difficult to attribute these new investigations to the efforts of the CTM leaders, since their earlier demands for action on the proposed reforms had gone unheeded.

The determining factor seems to have been the immminent presidential campaign of López Mateos.[35]

The opinions expressed to the congressional committees by representatives of labor and business groups in 1957 were basically the same as those expressed in 1952. Furthermore, the rivalry among the labor organizations that was evident during the 1952 hearings was still present in 1957. The CROM, for example, attributed the opening of the new hearings to CTM opportunism provoked by the forthcoming presidential campaign. As a result, the CROM showed less enthusiasm for profit sharing and stressed the need for "slow and serene" study of the matter. When the proposed CTM reforms were not reported out following the hearings,[36] the CROM affirmed that Congress "had succeeded in stopping the CTM pretensions dead in their tracks" and noted the CROM's obligation to prevent "other labor organizations from using the profit-sharing issue as a pretext for demagogic behavior."[37]

No congressional hearings concerning the CTM proposals were initiated after 1957. For the next few years, the CTM leaders continued to call for profit sharing at their National Council meetings. In addition, their representatives to the 1960 and 1961 National Assemblies of Labor Law presented petitions for profit sharing. Instead of specifying that the workers were to receive 10 percent of the profits, however, the 1960 petition asked that amendments to the Constitution and the Federal Labor Law specify the minimum percentage of profits to be distributed among the workers. The 1961 petition called for the establishment of special municipal commissions to decide the share of profits to which workers would be entitled.[38] The CROM, in the meantime, repudiated its original idea of obligating employers to finance a social security system for workers from their profits, since it had not been well received by other labor organizations, and began to consider elaborating a new profit-sharing project.[39]

In 1961, the year in which President López Mateos sent the reforms to Congress that were to pave the way for the establishment of profit sharing, the issue suddenly began to draw a great deal of attention. At the National Assembly of Labor Law held in November 1961, for example, there was a special committee devoted entirely to profit sharing, and eight petitions dealing with profit sharing were presented.[40] During the 1960 assembly, in contrast, only the CTM petition had been presented, and there had been no committee specifically devoted to the profit-sharing issue. Such sudden concern for the issue was no doubt the result of private indications by the government to members of the Academy of Labor Law, the sponsor of the assembly, that a special examination might prove useful.

The only public hint that the government was planning to establish profit sharing came from Salomón González Blanco, the secretary of labor. On June 21, 1961, upon leaving a meeting with the president, González mentioned that the day the profit-sharing provisions of the Constitution would be implemented was not far off.[41] The CTM, in its newspaper *Ceteme*, headlined the story reporting the secretary's remarks as a "formal promise of the administration" and reiterated its demand for profit sharing. Five months later, at a meeting of the CTM leaders in November, Fidel Velázquez, the head of the CTM, attempted to mobilize his subordinates and their rank and file in support of profit sharing, stating that the labor authorities were in a receptive mood and petitions on the part of labor groups would be welcomed. He urged the CTM regional leaders to exert pressure on the government through meetings, messages, public acts, and similar activities, so that the government would grant, among other things, the CTM demand for the establishment of profit sharing.[42]

When President López Mateos sent his proposed amendment to Article 123 of the Constitution to Congress in December 1961, the CTM leadership was able to claim that the president's action was in response to its demands. This claim no doubt gained the CTM leaders increased respect from their followers, who knew that their leaders had been asking for the establishment of profit sharing for a decade.

The fact that profit sharing had been demanded by the CTM leadership for many years had probably played a role in López Mateos's decision, because it enabled him to justify his reforms as a response to "worker aspirations" and thereby to reinforce the legitimacy of his decision. To consider the president's action a result of CTM demands, however, would be inaccurate. The CTM obviously had not been able to exert sufficient pressure upon the government during the preceding decade. Presidents Alemán and Ruiz Cortines had ignored its demands. López Mateos also could have ignored them, for by 1961 organized labor was not much more united in favor of profit sharing than it had been in 1951; profit sharing still ranked fairly low on the labor leaders' list of priorities and ranked even lower on the list of priorities of the rank-and-file members; and the CTM, the only labor organization that had expressed a demand for profit sharing with any fervor or consistency, was less powerful relative to other labor organizations in 1961 than it had been in 1951. In addition, if the profit-sharing decision had actually been made in response to worker demands, it would not have been necessary for the CTM leaders to attempt to mobilize their rank and file in support of profit sharing when they learned that the government might institute labor reforms. Thus, President López Mateos's 1961 decision to implement the profit-sharing provisions of the Constitution cannot be regarded as having been motivated principally by labor demands.[43]

The Ambivalence of the PRI

If the leaders of the organized labor movement were not vociferously demanding the establishment of a profit-sharing system and were, in fact, divided among themselves on the issue, other politically relevant groups showed even less interest and enthusiasm. The PRI took no clear stand because opinion within the labor sector of the party was extremely divided. The other sectors refrained from opposing profit sharing because it was one of the guarantees of the 1917 Constitution, but neither did they support the efforts of fellow party members to achieve it.

The issue of profit sharing was first raised at a PRI meeting during the party's 1951 National Assembly. A number of petitions were presented by various groups, and the party leaders decided to "recommend that the party's presidential candidate [Ruiz Cortines] take the proposals of the agrarian, labor, and popular sectors into account when formulating his program."[44] Among the proposals cited were those of the CTM and the Miners' Union for the establishment of profit sharing.[45] A specific endorsement by other PRI members, however, was conspicuously lacking.

Two years later, in 1953, the president of the PRI, General Leyva Velázquez, decided to comment on the issue of profit sharing since a congressional inquiry into the subject was in progress. His remarks could in no way be interpreted as an enthusiastic endorsement of profit sharing or of the CTM's specific proposals. He stated that the PRI "has looked with interest upon the efforts being made to have the profit-sharing provisions of the Constitution enforced" and added that it "is necessary to study and resolve everything related to the constitutional reform, without this being taken to mean that profit sharing should be established immediately." The president concluded by stating the necessity of an additional broad study before profit sharing could be introduced in Mexico.[46]

Nothing more was heard of the profit-sharing issue until 1960 when the PRI held its Third Regular Assembly. This particular assembly devoted itself to the task of drafting a new set of statutes, declaration of principles, and program of action for the party. Numerous proposals were presented by the various groups and the CTM leaders once again used the opportunity to call for the implementation of the profit-sharing provisions of Article 123 of the Constitution. This time the efforts of the CTM leaders met with partial success, for the Declaration of Principles that was elaborated by the PRI leadership and eventually approved by the delegates spoke of the workers' right to share in the profits of industry for the first time. The declaration stated that "the party demands that the effort necessary to find the most adequate system be made, in order that the workers share in the profits of industry, as is mandated in the Constitution."[47]

The CTM, however, had also insisted that a reference to profit sharing be incorporated not only into the Declaration of Principles but also into the PRI's Program of Action.[48] The latter consists of specific proposals that relate to the general principles expounded in the Declaration of Principles. The incorporation of the profit-sharing demand into both the Declaration of Principles and the Program of Action would have indicated that it was higher in priority than demands mentioned only in the Declaration of Principles,[49] and that there was general agreement about the way in which the principle should be realized. The PRI leadership did include a statement in the Program of Action to the effect that the party would "fight for the most perfect compliance with all the rights elaborated in Article 123," so in a sense profit sharing, a constitutional right, was covered by this declaration. However, the declaration also covered minimum wages, yet they were dealt with separately in another clause of the Program of Action.[50] One must therefore conclude that, by 1960, opposition to profit sharing had decreased and many (although not all) groups within the PRI were finally able to support it in principle, in spite of the fact that there was little agreement regarding the type of system desired.

The decreased opposition to profit sharing on the part of some labor leaders was a result of the less militant attitude of the Mexican labor movement in general. The section of the 1960 Declaration of Principles dealing specifically with labor matters, for example, no longer spoke of the class struggle (*lucha de clases*), as it had in the early 1950s, but of "class objectives."[51] Yet the earlier attitude did not die easily. For example, in an unprecedented 1961 television debate between representatives of the PRI and the PAN Lic. Antonio Vargas MacDonald, a well-known PRI leader who represented the party in the debate, was asked why the profit-sharing provisions of the Constitution had not been implemented. He replied, "Profit-sharing has been combated not by the government nor by bad leaders, but by revolutionary and resistant unionism that does not want to form associative relationships with employers because it proclaims the class struggle as an inherent and inevitable element of capitalism." The PRI spokesman left little doubt that he shared the attitude of these revolutionary labor leaders.[52] Thus, one year after the PRI had incorporated the demand for profit sharing into its Declaration of Principles and five months before President López Mateos was to send his profit-sharing amendment to Congress, the PRI leadership obviously remained ambivalent about the issue.

PAN Support for Profit Sharing

Unlike the PRI, the PAN unequivocally favored the introduction of a profit-sharing system in Mexico. The PAN abhorred the idea of a class struggle and instead stressed collaboration between workers and their employers, claiming that such collaboration was in the interest of both parties.

The PAN's Declaration of Principles and Program of Action did not contain any reference to profit sharing. The issue was first raised in 1952, when the PAN began to enter its own candidates in presidential elections. The platform of its first presidential candidate, Efraín González Luna, included a pledge to "work for the study and adoption of systems that would enable the workers to share in the profits of industry in order to achieve solidarity and collaboration between employers and workers, which is in their common interest, and the ultimate goal of social peace and an increase in production."[53] The same pledge was included in the platform of the PAN's 1958 presidential candidate, Luis H. Álvarez.[54]

In 1960 the PAN decided to act on its platform pledges. At its convention in September, Dr. Juan Landerreche Obregón, one of the party's founders, presented, at the request of José Torres, the party's presidential candidate in 1964, a *ponencia* (proposal) that outlined the specific profit-sharing system advocated by PAN.[55] The event received front-page coverage in Mexico's largest newspaper, *Excelsior*, which headlined the article: "The PAN will work for the reform of Article 123. It wants to make profit sharing effective."[56] The system proposed by the PAN was similar in many aspects to the system eventually adopted by the National Profit-Sharing Commission in 1963. It provided for a system based on profits, not salary, to be determined on the basis of income-tax declarations. Workers were to share only in profits and not in losses. Intervention in the administration of a business enterprise was to be expressly forbidden. Workers were to be given the right to object to the amount declared as profits by their employers. In such cases, the federal government would determine the validity of the objection. The municipal commissions that had previously had the power to establish municipal profit-sharing systems were to be suppressed. The main difference between the system proposed by the PAN and the system ultimately adopted by the Mexican government was that the former allowed employers to deduct a certain percentage of their invested capital from profits before distributing anything to workers, while the latter did not allow that type of deduction.[57]

The PAN never presented its profit-sharing proposal to Congress, mainly because the party was not equipped to undertake the detailed studies of the economy that were necessary before a specific profit-sharing system could be adopted. The PAN thus felt that it could do nothing more; the establishment of a profit-sharing system had to await positive government action.[58] As a result of its inaction in Congress, however, the PAN lost an opportunity for a propaganda coup at the expense of the PRI. The PAN profit-sharing proposal clearly would have been defeated, since the PRI deputies would not have voted in favor of a PAN proposal that would have its opponent in the position of implementing a provision of the revolutionary Constitution.

In the 1964 campaign for election to the Chamber of Deputies, one PAN candidate used the following poster: "Profit-Sharing: The Constitution provided for it in 1917; the PAN asked for it in 1960; the government established it in 1961; employers, laborers and the government approved it in 1963. Social justice is the task of all." The candidate was obviously implying that the PAN's demand for profit sharing had played an important role in President López Mateos's decision. To the public, this assertion seemed plausible, for the fact that López Mateos had already taken the decision to establish a profit-sharing system was still a well-guarded secret. The PAN did not know of the president's decision at the time of its convention in September 1960. The PAN's demand for profit sharing cannot, however, be considered the cause of López Mateos's decision. The PAN, like some elements within the PRI, had first called for profit-sharing in the very early 1950s, and no government action had resulted. If López Mateos had not wanted to establish a profit-sharing system, neither the arguments of the PAN or the PRI would have induced him to do so.

Probably the main effects of the PAN's demand for profit sharing were embarrassment on the part of the government and confusion on the part of the general public. The 1960 PAN demand was particularly disconcerting, for the labor movement had lost some of its earlier radicalism and as a result, there was less opposition to the idea of profit sharing than there had been in the early 1950s. The government was embarrassed because a so-called opposition party was calling for the implementation of provisions of the revolutionary 1917 Constitution that revolutionary governments had failed to enforce. The public was confused because the PAN, consistently and pejoratively labeled "conservative" or "reactionary" by progovernment interests, was calling for the enforcement of provisions of the revolutionary Constitution. Thus, either the PAN was not that conservative or profit sharing was not that revolutionary or both. The embarrassment and confusion may have served López Mateos well, however, for he may have persuaded some of the less enthusiastic labor leaders that the establishment of profit sharing could no longer be postponed now that the PAN was demanding it.

Opposition from the Private Sector

While the organized labor movement, which supposedly would benefit from the establishment of profit sharing, remained divided and generally unenthusiastic, and the official party continued ambivalent, the private sector remained relatively unified in its opposition to an obligatory profit-sharing system. Such opposition, however, could not be openly expressed because profit sharing was provided for by the 1917 Constitution and to oppose it openly would make the private sector vulnerable to charges that it was "unrevolutionary." As a result, the efforts of the private sector were directed toward postponing the establishment of profit sharing or toward substituting for the workers' right to profit sharing some other benefit that the private sector considered less detrimental to its interests.

The first public indication of the sentiments of the private sector on this issue occurred during the First Congress of Industrial Law in 1934. The business delegates did not attack the concept of profit sharing, but voted for a social security system for workers to be financed by the employers from their profits. This system was designed to replace the workers' constitutional right to profit sharing.[59] The Mexican government never acted on this vote.

In 1951, the private sector used the National Convention of COPARMEX as a forum in which to express its views on the profit-sharing proposals that the CTM had just submitted to Congress. The opinions expressed at this convention were later sent by CONCAMIN, CONCANACO, COPARMEX, and the Bankers' Association to the congressional committee conducting an inquiry on the issue of profit sharing. In essence, the private sector claimed that profit sharing required a spirit of cooperation between workers and their employers that did not yet exist in Mexico. In its absence, profit sharing would only increase labor-patrón conflict, because the workers would use it for disruptive behavior. The business leaders also alleged that obligatory profit-sharing systems had failed in almost all countries where they had been introduced, and they contended that there was thus no reason to assume that such a system would be successful in Mexico. The private sector further asserted that the best way to raise real incomes was not by means of a profit-sharing system, but through the expansion of production. Such expansion required private capital investment, and profit sharing would decrease the amount of investment by the private sector. To increase production it would be necessary to provide incentives for the workers, and the private sector claimed that higher salaries, for example, would constitute a more efficient incentive than profit sharing. The business groups warned that if a profit-sharing system were introduced and if it were to function poorly (which was to be expected!), Mexico's industrialization and economic development might suffer.[60] These views were repeated six years later to the congressional committees that held closed hearings on the profit-sharing issue in 1957. The private sector's objections to an obligatory profit-sharing system received a final reaffirmation in 1961 in a study commissioned by CONCAMIN.[61]

The attitudes and opinions just discussed were essentially those of the organized groups that are the official representatives of the private sector. Some businessmen, however, did favor the establishment of a profit-sharing system. They were mainly Catholics, who based their support for such a system on the progressive papal encyclicals *Rerum Novarum* and *Quadragesimo Anno*. At the 1960 meeting of the Catholic-oriented Social Union of Businessmen, for example, the main speaker stated that "as Catholics and as Mexicans," businessmen should share their profits with their workers, since that was one way of increasing social justice. The specific profit-sharing system endorsed by this group consisted of a fund to be financed by employers from a percentage of their profits. A worker was to be entitled to receive money from this fund only when he left his job or was confronted by an emergency. If he died, his family would receive money from it.[62]

Although some businessmen favored profit sharing, the overwhelming majority were opposed to such a system, whether it was to be obligatory or voluntary. As a prominent panista pointed out, the proof that Mexico's private sector did not really favor profit sharing lay in the fact that a very negligible number of employers voluntarily established it, even though profit sharing was a right provided for by the Constitution.[63]

The Reasons for the President's Profit-Sharing Decision

The absence of strong demands for profit sharing on the part of the sector it was supposed to benefit, the expression of opposition by the groups whose profits were to be reduced, and the relative neutrality of other groups in society suggest the need to search elsewhere for an explanation of the decision of President López Mateos to implement the profit-sharing provisions of the Constitution. The answer lies in an examination of the president's background, interests, and political beliefs; the then-prevailing political and intellectual climate; and Mexico's level of development.[64]

Adolfo López Mateos was the first Mexican who had served as secretary of labor immediately prior to becoming president,[65] and throughout his term he consistently tried to identify with the workers. His words to the National Council of CTM members in 1960 are typical of his public attitude toward labor: "Inform the workers . . . that there exists an absolute identification between my government and the labor movement that the CTM represents."[66]

The president's prolabor orientation was evident not only from his words but also from his actions. His 1961 reforms were not the only evidence of his concern for labor's well-being. In 1959 he reformed the Social Security Law so that it would include part-time workers, and in the following year he amended Article 123 so that government employees would be protected by minimum-wage legislation.

López Mateos had wanted to be the president who would completely reform the Federal Labor Law, which had been enacted in 1931. Soon after he took office, he named a three-member commission to begin work on the Article 123 and Federal Labor Law reforms. He soon discovered, however, that there was insufficient time to complete the enormous amount of work involved so he abandoned his idea of a total reform.[67]

Still another indication of López Mateos's favorable attitude toward organized labor were his efforts to slow the increase in the cost of living. He substantially expanded the government's Compañía Nacional de Subsistencias Populares (CONASUPO) mobile unit program, a program that distributed high-quality staple goods to workers and peasants at prices significantly lower than prevailing market prices. And as a result of sharp increases in the minimum wage and tight control of prices during his administration, the workers' real minimum wage index reached its highest point since the presidency of Cárdenas. In fact, between 1942 and 1960, with the exception of the 1952-1953 fiscal year, the real minimum wage index consistently began to exceed its 1940 level.[68]

The relatively lower real minimum wages that prevailed throughout the 1940s caused the normally docile and unmobilized labor rank and file to become increasingly militant, and in the mid

1950s it began to repudiate its "do-nothing" leaders. Several unions then came under the control of new, more radical leaders who led their followers into a series of strikes in the period of 1957 to 1959. First the telegraphers struck, followed by the left-wing members of the primary-school teachers in the Federal District. Then several important locals of the petroleum workers went on strike. The railroad workers precipitated the most disruptive strike, however, leaving thousands of people stranded during the 1959 Easter vacation. The government suppressed the strike and arrested large numbers of railroad workers and their radical leaders. The military occupied the union's offices, and soldiers took charge of running the trains.

It is not particularly surprising that López Mateos was confronted with so much labor unrest immediately upon assuming office. As already noted, the strike in Mexico is frequently a political weapon, and it is used most often against presidents considered sympathetic toward labor. The 1957-1959 strikes were atypical in that they were not led by co-opted leaders who were sympathetic to the regime. The deteriorating economic situation of organized labor had mobilized the normally unmobilized rank-and-file members, who proceeded to repudiate their co-opted leaders and replace them with more radical ones. González Casanova notes that this kind of behavior occurs infrequently.[69] The timing of the strikes, however, was typical, since they took place during the regime of a prolabor president.

The president's actions during the railroad strike greatly damaged the prolabor image that he had cultivated prior to becoming president and that he had hoped to strengthen during his term in office. He succeeded in refurbishing his image through his efforts on behalf of the labor movement after the 1957-1959 unrest. In fact, it is probable that López Mateos's personal desire to benefit labor was both augmented and made easier to realize precisely *because* of the high degree of mobilization of the labor movement immediately preceding and continuing into the first year of his term of office. It caused the president to place satisfying the labor movement even higher on his list of priorities and enabled him to justify a more extreme prolabor (and perhaps antibusiness) stance as both necessary and just in view of the long period of deprivation and sacrifice suffered by the labor movement prior to his administration.

In addition to his labor background and positive identification with the labor movement, López Mateos was considered to hold more liberal political beliefs than his three immediate predecessors. His liberal image was reinforced by his celebrated Guaymas speech of July 1, 1960, in which he declared that his government was "within the Constitution, of the extreme Left."[70] It was further enhanced by the changes he made in the composition of public investment. Social-welfare investments reached a historical peak of 19.2 percent of total investment,[71] while investments in infrastructure substantially declined.

In sum, it is understandable that López Mateos, a "left of center" president and a former secretary of labor who made a special effort to identify himself with the working class, would, in his desire to create a historical image, consider some project that would involve the labor movement. The implementation of certain dormant provisions of Article 123 of the Constitution represented the optimal opportunity for López Mateos to accomplish his goal. First, almost all Mexican presidents seek such an opportunity, because it symbolizes that the Mexican Revolution is an ongoing process and is not dead as some of its critics have claimed.[72] This consideration may have been important in López Mateos's decision to implement the profit-sharing provisions, since the 1959 Cuban Revolution had made Mexico seem rather unrevolutionary in comparison with Cuba. Second, implementing any provision in the Constitution reinforces the "revolutionary" consensus of the Mexican regime and indirectly, the legitimacy of the president by more closely identifying him with the revolution that brought to power the political elite to which he belongs. In fact, the implementation of provisions of the Constitution has become a kind of moral imperative for Mexican presidents. Third, the implementation of a provision of Article 123 of the Constitution appeared to be a more significant decision than it might prove to be in reality. In effect, it enabled the president to increase his support among members of the labor movement and thus keep them demobilized by means of a pay-off that was largely symbolic, thereby conserving the regime's tangible or material resources for other purposes.[73]

Although Article 123 contained numerous sections that had never been adequately enforced, López Mateos had earlier expressed his support for implementation of the profit-sharing clause. In a 1953 meeting, during which representatives of the CTM had asked for a profit-sharing system, Secretary of Labor López Mateos had stated that the CTM profit-sharing petition, "in the sense that it would give force to the constitutional provision for profit-sharing, merits the president's approval, considering that what is involved is a mandate of our Magna Carta, which therefore should be respected by making it the object of adequate implementation."[74]

The fact that a profit-sharing system would contribute toward a redistribution of income provided the president with an additional incentive. Since the late 1950s, a number of outstanding Mexican economists had been criticizing the inequitable distribution of income that existed in Mexico on social and economic grounds.[75] They charged that it was a perversion of the goals of the Mexican Revolution. They also argued that the country's economic growth would slacken unless the domestic market were expanded by increasing the purchasing power of the masses.[76] Several prominent members of the private sector supported the latter argument because they believed that the insufficient consumer demand prohibited many industrial enterprises from operating at full capacity.[77]

Although the introduction of a profit-sharing system was one way of effecting the redistribution of income called for by a segment of the intellectual elite, it was not the best way. Profit sharing would essentially remove some money from the private sector and give it to organized labor, a group that was far from being the most economically depressed and discriminated against in Mexico. If anything, organized labor was, relatively speaking, in a fairly good position vis-à-vis rural laborers and nonunionized urban masses.

Changing the tax laws might have been a better way of effecting a redistribution of income, since these laws were somewhat regressive. The existing laws relied rather heavily on indirect taxes and favored persons with high nonwage incomes.[78] In fact, tax reform was unanimously recommended by the intellectuals who criticized the inequitable income distribution. (None of these critics suggested instituting a profit-sharing system as a solution to the problem.) Raising tax rates on larger incomes and increasing the total amount of money collected would provide the government with additional resources to raise the standard of living of the masses.[79]

López Mateos did not, however, make any radical changes in the tax system.[80] He did not necessarily reject this option because he thought tax reform would have encountered more opposition than the establishment of profit sharing. There were no grounds for assuming that the profit-sharing decision would be unopposed. In fact, the profit-sharing reforms engendered a substantial amount of opposition.[81] It seems more likely that the government refrained from drastically changing the tax laws in the belief that the existing system was helping the country attract the foreign and domestic private investments encouraged and solicited by the political elite.

López Mateos therefore did not make the profit-sharing decision principally to make a substantial redistribution of income in favor of the lower classes. His principal interests were to enhance his historical image and to do something for the organized labor movement in order to demobilize it and to refurbish his image as a president identified with organized labor. The fact that the president's search for an appropriate vehicle coincided with a period during which a segment of the intellectual elite was advocating a redistribution of the national income provided an additional incentive for his selection of the profit-sharing clause, for it enabled him to appear to correct a state of affairs that had been much criticized by a segment of the intellectual elite without actually doing so.

López Mateos's predecessors had also considered implementing the profit-sharing provisions. Why had they refrained from doing so? A brief examination of the technical requirements for a successful profit-sharing system indicates that Mexico's low level of industrial development has until fairly recently precluded the possibility of establishing a well-functioning and enforceable system in that country.

In order for an obligatory profit-sharing system to function well, accurate information regarding profits must be available to the government. Income-tax declarations are the best source of such information. In 1917, when profit-sharing was made a constitutional provision, the government had no way of knowing who, if anyone, was obtaining profits because there was no income-tax law. The first such law was passed in 1921; however, it provided only for a temporary system, to be reestablished each year. Only in 1924, when the first permanent income-tax system came into being, did businesses begin to keep records for the government.[82] Another early obstacle to the establishment of profit sharing was the fact that Mexico had very little industry until the administration of Miguel Alemán (1946-1952). Beginning with the late 1940s, industrial development began in earnest, and by the mid-1950s there was little doubt that Rostow's famous "take-off" had occurred. It would have made little sense to attempt to establish a profit-sharing system at a time when industrialization was incipient and still precarious.

A third problem involved the availability of accurate statistical information about the Mexican economy. Such information is desirable in order to judge or estimate profit sharing's effect on income distribution, reinvestment, consumer demand, and economic growth in general. These data, however, have been available at least since the 1950s, and perhaps earlier.

Although these early problems had been solved by the time López Mateos took office, there were other barriers to be overcome. One problem involved the control the government was able to exercise over the reporting of profits. If profits were not being reported accurately on income-tax returns, then opportunities for depriving the workers of the percentage of profits legally due them would be abundant. Past presidents considered implementing the profit-sharing provisions, but when they had discussed the possibility with economists and lawyers, they had been told that adequate control over the reporting of profits had not yet been achieved.[83] This problem had not been definitively resolved by the time López Mateos took office. It was evidently of great concern to the incoming president, for during his campaign he asked Hugo Margáin, who was then head of the Tax Bureau, whether existing fiscal controls would permit profit sharing to be successful. Margáin replied that he was completely satisfied with government control with regard to the big firms. In the medium-sized firms (those capitalized between 1 and 5 million pesos) he felt that control was satisfactory in 75 to 80 percent of the cases. He stated, however, that control was inadequate in small firms with a total capitalization of under 1 million pesos, because in these firms elaborate accounting procedures were too costly and therefore unfeasible.[84] During my interview with him, Margáin left little doubt that President López Mateos would not have decided to institute a profit-sharing system had he thought that the government would have been unable to enforce it.[85]

The government's ability to exact compliance with tax laws was related to its ability to enforce the private sector's compliance with laws in general. There always existed at least a theoretical possibility that the private sector would in some way seek retaliation against a government that was insisting on implementing an unwanted profit-sharing system. Whatever this retaliation might be, it was conceivable that it could prove extremely damaging to the Mexican economy. As was noted earlier, however, the federal government has for some time had more influence over the private sector than the private sector has had over the government. With the passage of time, the possibility of serious economic damage as a result of private-sector opposition to an unpleasant government-established system such as profit-sharing has been continually decreasing.

A final obstacle to early implementation of the Constitution's profit-sharing provisions related to the system's possible effect on worker-employer relations. There is a widespread belief that profit sharing contributes toward improved relations between the workers and employers because it gives the workers a greater stake in the enterprise. As a result, the workers will presumably act more responsibly and "rationally" (i.e., they will avoid conflict) in order not to cause the firm to lose profits, which ultimately would decrease the amount of money they receive. This argument does not apply, however, to cases in which laborers and their employers are continually in disagreement. In such cases, the workers' distrust of their employers can cause them to challenge the accuracy of reported profits, thereby expanding the arena of conflict between employers and employees.[86] López Mateos was aware of this problem, for he reportedly stated privately that he did not consider profit sharing a good mechanism for resolving labor-business conflicts.[87] This remark implies that he would only have established a profit-sharing system when relations between labor and business were relatively free from strife.

The relatively good working relationship between business and labor that is necessary for a successful profit-sharing system is of recent origin in Mexico. In the early 1950s, for example, when some private-sector leaders called for business-labor unity, the CTM responded that the labor movement was not yet in a position to collaborate since there were still many nonunionized and hungry workers and many unanswered labor demands.[88] At a 1952 CTM convention, the head of the CTM stated that "the social justice that the employers understand is that of violating the rights of the working class."[89] By the early 1960s the situation had improved greatly. In May 1964, for example, a group of labor leaders for the first time accepted an invitation to attend a convention of COPARMEX Employers' Centers.[90] By the time López Mateos took office, therefore, labor-business conflict had subsided to the point where it no longer was an insurmountable obstacle to successful implementation of the profit-sharing provisions of the Constitution.[91]

López Mateos was thus one of the first presidents who served at a time when the establishment of a viable profit-sharing system was a real possibility. Technical feasibility, however, was not a sufficient cause. Also required was a president with a special background and a definite commitment to the labor movement. López Mateos was such a president.

The preceding analysis does not imply that a Mexican president would only decide to establish profit sharing if he were certain that a profit-sharing system could function. Ostensibly bold executive decisions are often made when there is no possibility, or perhaps no intention, of enforcing them. Such decisions are frequently characterized as demagogic or romantic. With respect to profit sharing, however, López Mateos had limited options. The demagogic or romantic decision already had been made by the members of the 1916-1917 Constitutional Convention. Profit sharing had been a "paper law" since 1917. The only possibility for presidential action in this area, therefore, was to implement an already existing law.

Government Initiation and 4
Interest-Group Response

It is impossible to state exactly when President López Mateos became committed to the implementation of the profit-sharing provisions of Article 123. During his presidential campaign (late 1957 and early 1958), he indicated to Hugo B. Margáin that he was seriously considering establishing a profit-sharing system. By mid-1959, following the railroad strike, he had apparently concluded that significant and visible benefits would have to be given to organized labor if future unrest was to be avoided. It therefore appears that López Mateos committed himself to the establishment of profit sharing sometime between late 1957 and the middle of 1959.

López Mateos discussed the idea with a number of high-ranking individuals in the Treasury, the Ministries of Agriculture, Labor, and Industry and Commerce, and several government banks, especially the Bank of Mexico. "Discretion was recommended," which meant that information about what was occurring was to spread no further.[1] The deliberations were intentionally kept a well-guarded secret.

Apparently there was little dissension among the individuals involved in the discussions, and the general agreement encouraged López Mateos to proceed with his idea of implementing the constitutional provisions.[2] He therefore appointed a three-member *ad hoc* commission[3] and assigned it the task of drafting a constitutional amendment to elaborate the general guidelines of the profit-sharing system. The amendment was also to include additional reforms beneficial to organized labor. The commission began functioning early in 1960.[4]

The individuals who participated in the earlier deliberations had decided that the establishment of a profit-sharing system would not be welcomed by all groups and that the private sector would be especially reluctant to embrace it. The creation of the three-member commission and the nature of its task were therefore not announced. It should be pointed out, however, that secrecy in the drafting of labor legislation is not necessarily a general rule in Mexico. In some instances the president or his ministers inform the public that labor legislation is being drafted and opinions are solicited *prior* to sending the proposed legislation to Congress. In 1968, for example, the Díaz Ordaz government sent copies of a draft of a new labor law to interested parties for their comments. It appears that secrecy is employed only when it is expedient, that is, when the president seeks to avoid the mobilization of potential opponents of a controversial piece of legislation. Unexpected reforms encounter unprepared critics.[5]

The secrecy surrounding the profit-sharing decision was compounded by the president's deliberate avoidance of any public mention of the issue. His campaign speeches, his May Day speeches (addressed specifically to the labor movement), other speeches delivered before labor groups, and his state of the union messages made no reference to the impending establishment of profit sharing. The most telling hint of the president's intentions came in June 1961, when his secretary of labor stated that "Clauses VI and IX of [Article 123] of the Constitution that provide for [profit-sharing] have no implementing legislation, which has made it impossible to enforce them, but the day is not far off when legislation to implement the profit-sharing provisions of the Constitution will be passed."[6] The campaign speeches of López Mateos did mention the need to increase the purchasing power of the masses, the need to apply unapplied labor laws, and the right of workers to share fairly in the fruits of their labors.[7] In the message he delivered upon assuming office, he spoke of the need for a more equitable distribution of income.[8] Every president's campaign speeches, however, touch upon these themes, and there was little reason to assume that more than rhetoric was involved in this case.

Demobilizaton of Opponents and Mobilization of Support

The profit-sharing reform remained a well-guarded secret until December 27, 1961, when the proposed amendment was sent to Congress for its approval. The private sector was taken by surprise. Not even its leaders had known about the impending legislation. When he was asked his opinion of the presidential initiative immediately after the reforms became public, Lic. Campillo Sainz of CONCAMIN stated, "We know nothing more about the presidential initiative than what was published in the newspapers."[9] A COPARMEX spokesman also said his organization was unable to give an opinion because it was not familiar with the project.[10] A paid COPARMEX advertisement "profoundly lament[ed] the procedure . . . of sending to Congress a constitutional reform project without previously sounding out the authorized opinion of the employer sector, whose interests are involved in the cited project . . ."[11]

Some leaders of the organized labor movement claimed that they knew in advance about the López Mateos reforms. However, they probably knew only that the president was going to do *something* to benefit the labor movement. It is unlikely they were sure of the specifics of the proposed legislation. One of the labor representatives on the 1963 National Profit-Sharing Commission, for example, stated that the reforms were not a surprise "since labor knew that President López Mateos would pay attention to the demands of the workers, and the workers had been fighting for profit sharing for a long time."[12] A CTM lawyer spoke of the labor movement's "being the first to know of the favorable attitude [*buena disposición*] of the president toward what labor wanted."[13] Fidel Velázquez (the head of the CTM) and perhaps one or two other important labor leaders had obviously been given some indication that significant labor reforms were about to be made. As noted earlier, one month before the reforms were sent to Congress, Velázquez announced that at a meeting of CTM officers, "it was agreed that all the CTM organizations should express, in every way possible . . . messages, manifestations, public acts, meetings, etc., to the Congress, the government authorities, especially the president, the interest that the Mexican workers, represented by the CTM, have in [CTM proposed reforms to Clauses III, VI, IX, XXI and XXII of Article 123]." (These reforms included the establishment of profit sharing.) Velázquez added that "the labor authorities are in the best mood for attending to the petitions . . ."[14] It thus seems that the president or the secretary of labor privately had informed the principal leaders of the Mexican labor movement that some important reforms were forthcoming, so that the labor leaders could mobilize their rank-and-file membership to petition the government for such reforms. This maneuver would create the impression that the presidential amendment was a direct response to pressure from the labor movement, even though it was not.

The content of the presidential reforms, like the secrecy surrounding their drafting, was intended to avoid the mobilization of potential opponents. The provisions of the amendment were nonspecific and at times extremely vague, so there were few concrete provisions to which one might object. The profit-sharing amendment essentially called for the creation of an *ad hoc* national commission that would undertake the necessary economic investigations and then elaborate a profit-sharing system. The legislation itself did not propose a specific system nor did it deal in any way with the percentage of profits that would be distributed to the workers. Furthermore, the amendment proposed a tripartite commission, so that business and labor interests knew that they would eventually have an opportunity to express their views and perhaps exert a decisive influence upon the ultimate outcome.

The government had also taken care to assuage the principal fears of both business and labor. In the subsequent Senate debates on the amendments, a senator noted the government's efforts in this direction, calling the reforms "a happy formula" designed to "calm all those elements that consider themselves affected by the laws . . ."[15] In an attempt to forestall objections from the private sector, the amendment provided for studies of the national economy that were to precede the determination of the type of profit-sharing system to be adopted. It stated that the decision should take into consideration the need to foster the industrial development of Mexico, the reasonable return to which capital is entitled, and the necessary reinvestment of capital.[16] It exempted new industries from profit sharing for a number of years. Most important, it stated categorically that "the right of workers to share in profits does not imply the ability [*facultad*] to intervene in the direction or the administration of the enterprises."[17]

The fears of the leaders of organized labor that the private sector would not report its profits truthfully were allayed by the provision that the workers would have the right to object to the amount of profits reported by their employers on their annual tax returns and to ask the government to verify the accuracy of the figure. The percentage of profits received by the workers was to be revised by a national commission every ten years and was not to be connected in any way with bargaining every two years over minimum-wage rates and collective contract provisions. This stipulation answered the objections of the more left-wing leaders who had contended that what the workers gained in profits they would lose in wages and fringe benefits.

Timing also played a role in keeping mobilization of opposition to a minimum. The constitutional amendment was sent to the Senate on December 27, 1961, several days before the end of that year's session of Congress. It was unanimously passed the following day, after a short "debate" (the record filled a mere eight pages) that consisted mainly of laudatory remarks about the legislation and President López Mateos. The amendment was then sent to the Chamber of Deputies, where, on December 29, 1961, it also received unanimous approval and lavish praise. Because Congress received the reforms at such a late date, there was not time to make even minor changes. As Deputy Gamboa Pascoe pointed out in the Chamber debates, "any addition . . . would impede the approval of this law by Congress in the present session, which is about to draw to a close."[18] Rather than delay a constitutional amendment proposed by the president until the next session of Congress nine months later, the deputies refrained from making several additions to the amendment that they might have made had time permitted.[19]

Another tactic the government employed to minimize mobilization of opponent forces was to make the amendment seem less special than it was. The government accomplished this end by deluging Congress more or less simultaneously with legislative proposals that would benefit other groups. By this action, the government lessened the possibility that its prolabor reforms would provoke dissatisfaction or feelings of unfair treatment and neglect on the part of the other groups, thus enhancing the president's ability to mobilize these nonlabor groups in support of his labor legislation. In December 1961, therefore, the month President López Mateos sent his amendment to Article 123 to Congress, he also sent reforms that provided for a social security program for the military, an agricultural insurance law for peasants, farmers, and cattlemen, and a new headquarters building for the government bureaucrats' union.

Prior to undertaking the whole series of labor reforms, government spokesmen frequently commended organized labor on its good behavior. In March 1961, for example, *El Nacional*, the government newspaper, devoted an editorial to "the good labor-business relations" and praised "the harmony that today exists between the workers and the entrepreneurs."[20] In May 1961, *El Nacional* ran a front-page story with a headline "A Tranquil Labor Panorama" in which it quoted the secretary of labor as saying that "no worker-employer conflict of importance presently exists in the country, and all [disagreements regarding contract] revisions are being resolved in a spirit of comprehension that eliminates all bitterness and enables them to be settled without going to extremes."[21] This praise of the absence of conflict was motivated by a desire to impress upon the labor movement that the president acts favorably toward it only if labor acts responsibly and refrains from resorting to violent or illegal means (such as the 1957-1959 strikes) to obtain what it seeks. As Albert Hirschman noted in his study of the Latin American policy-making process, "once it has become clear that policy-makers are responsive to threats of violence in one particular area, such threats will be delivered with increasing frequency."[22] By their behavior, the Mexican decision makers were indicating that little was to be gained by violence and a great deal could be expected by those who cooperated with the regime's leaders.

Following the passage of the 1961 constitutional amendment the president and his ministers went to great lengths to placate the private sector. The fact that the profit-sharing decision had been made without consulting or informing business interests had shocked, frightened, and angered the private-sector leaders. Relations between the government and the private sector had been deteriorating ever since López Mateos had assumed the presidency.

His nationalization of the electric-power industry in 1960 and his government's liberal attitude toward Cuba were among the contributing causes. The president's secret decision to institute a compulsory profit-sharing system added fuel to the fire. As early as 1960, the López Mateos government initiated an effort to pacify the private sector. There were speeches in which "the president and his principal lieutenants repeatedly emphasized the loyalty, respect, and support that the government was prepared to offer to Mexico's domestic businessmen."[23] For the first time representatives of the private sector were invited to accompany a Mexican president on an international tour. Numerous editorials began to appear in the government newspaper stressing that the government and the private sector were not in competition with each other, and praising the private sector for its increasing concern with social justice.[24] These efforts increased immediately preceding and following the announcement of the profit-sharing decision in December 1961.[25]

The government's public praise of the private sector was followed by a more substantial behind-the-scenes effort to repair its relationship with business leaders. It also wished to convince them that the upcoming profit-sharing system would not prove detrimental to their interests. This effort was particularly necessary in view of the fact that the private-sector leaders had reacted to the surprise constitutional amendment by joining together to plan a strategy to protect their interests. Part of this strategy involved mobilizing their subordinates to support their efforts.

The strategy that the López Mateos government employed to demobilize the private-sector leaders and win their support essentially involved incorporating them into all phases of the profit-sharing decision once the "decision in principle" (i.e. the implementation of the profit-sharing clauses of the Constitution) had been made. The private-sector leaders could not openly criticize the principle of profit sharing because it was included in Article 123 of the 1917 Constitution. They were, therefore, limited to criticizing substantive aspects of the profit-sharing decision. One of the targets of their criticism was the government's failure both to inform and consult them regarding the 1961 amendment. By incorporating the private-sector leaders into subsequent phases of the process, an important rationale for their opposition would be undermined.

The president was also willing to involve the leaders of the private sector in the next phase of the profit-sharing decision because the most controversial aspect, the decision in principle, had already been made. The next phase involved the drafting of reforms to the Federal Labor Law, which were necessary because the 1961 constitutional amendment had caused the Constitution and the Federal Labor Law to be incongruent. The labor law reforms thus involved decisions on specific, often technical, details rather than on broad ideological principles. The incorporation of the leaders of the private sector into the decision-making process during this next phase, therefore, would place them in the position of having approved and accepted the president's decision in principle.

In May 1962, the secretary of labor, in a private communication to the leaders of the principal organizations of the private sector, stated that the government was in the process of drafting reforms to the Federal Labor Law and invited the private sector organizations to name representatives to a committee that would "exchange impressions" with a group of government representatives.[26] The invitation was accepted. At the same time the secretary of labor made an equivalent invitation to the leaders of the organized labor movement,[27] who also accepted. The organized labor movement was the theoretical beneficiary of the profit-sharing decision. Furthermore, although the profit-sharing decisions had not been made in response to strong and consistent demands on the part of the labor leaders, most of them had gratefully accepted it. Some labor leaders, however, remained unenthusiastic even after the president's decision.[28] The purpose of the secretary of labor's invitation to the labor leaders was to mobilize them in support of the decision and to encourage them to mobilize their rank and file, most of whom had remained silent and passive after the decision was announced.[29]

The commission to which the representatives of the private sector and the organized labor movement presented their views in the middle of 1962 was the same three-member commission that had been formed in 1960 to draft the 1961 constitutional amendment. The government, however, did not make public either the existence of the commission or the fact that it was soliciting the viewpoints of labor and business leaders until October 1962.[30] The reason was that the 1961 constitutional amendment technically was not legally binding until it had been ratified by a majority of the

Mexican state legislatures, duly approved by the Congress, and signed by the president. The required approval by a majority of the state legislatures occurred in March 1962; however, no further action could be taken until the Chamber of Deputies and the Senate, which were in recess, reconvened in September 1962. At that time they reviewed the votes of the state legislatures and declared the amendment to Article 123 duly approved. Only then could President López Mateos announce the formation of the three-member commission to draft reforms to the Federal Labor Law. For the president to have revealed that the three-member commission had begun drafting the reforms prior to final approval of the 1961 constitutional amendment would have constituted an embarrassing public acknowledgment that he regarded the ratification by the state legislatures and the Congress as the formality it actually was.

It appears that the president and his advisers originally underestimated the amount of opposition that the labor law reforms might generate, for the secretary of labor had promised the business representatives that they would be shown a copy of the reforms before they were formally submitted to Congress for approval.[31] There was nothing unusual about such a promise, and the procedure was exactly the same as the one followed in 1968 when the Díaz Ordaz government sent copies of a draft of a completely new labor law to interested parties and solicited their opinions before sending the law to Congress. However, President López Mateos and his advisers eventually decided against divulging the contents of the 1962 labor law reforms before sending them to Congress. In a meeting with a three-man delegation representing CONCAMIN, CONCANACO, and COPARMEX on August 31, 1962, the secretary of labor vehemently denied that he had ever promised to show the draft of the labor reforms to representatives of the private sector prior to sending them to Congress. To do so, he said, would "diminish the stature of the President of the Republic, since only he could submit the reforms to Congress which represents the people." The secretary of labor then asked the private-sector representatives to consider the implications: "The proposed reforms would have to be circulated by the confederations, both to their committees as well as to their councils and to the respective chambers, in such a way that copies would even be made for consultation by third parties. In this way the labor sector would learn that the secretary of labor gave the project to the private sector, and with that, the workers could not be denied copies of the reforms. The reaction would be an uproar [*alarca*] in which at least a hundred people would feel the obligation or the desire to discuss and elevate the matter to a demagogic level, which would not be the best thing for the patrones, because there would be pressure on the government to take more from the private sector, using the proposed labor law reforms as a base."[32]

The government's solicitous concern for the private sector was undoubtedly only part of the explanation. More important was the probability that the mobilization of both business and labor groups would have decreased the government's decision-making autonomy and would have raised the level of conflict, thus placing undesirable stress on the regime. The president wished to avoid such unnecessary conflict and thus reversed his initial decision to show the proposed reforms to the interested parties before sending them to Congress. Following this reversal, the secretary of labor attempted to placate the business representatives with an offer to meet with them at a later date (but before the reforms were to be sent to Congress) to discuss which suggestions the government had accepted or rejected and why.[33]

The representatives of the private sector and those of the labor movement presented their respective views to the government commission and the secretary of labor separately. Thus labor and business focused their activities upon the government, specifically the Ministry of Labor, rather than upon each other. This arrangement reduced the possibility of contact between the conflicting parties, as well as the probability that either party would ever really discover exactly what views and opinions had been expressed by the other to the secretary of labor and the three-member commission. The executive thus had maximum leeway in preparing the 1962 reforms, since it could pick and choose among the opinions and ideas it received from each side and eventually claim that the resulting reforms represented a fair compromise among conflicting demands. Because of the mechanics of the deliberations, business and labor would find it difficult ever to ascertain whether or not the reform really represented a compromise.

The Federal Labor Law Reforms

On December 20, 1962, the president sent the proposed reforms to Congress. The section of the reforms relating to profit sharing elaborated the general guide-lines that the members of the soon-to-be-formed National Profit-Sharing Commission were to follow in their determination of the specifics of the new system. It stated, for example, that the profit-sharing system was to be based on income-tax declarations and that the share of profits each worker was to receive would be divided into two parts, one that was to be the same for all workers based on the number of days worked annually, and another that was to vary in direct proportion to a worker's wages. It also specified the institutions or industries that were to be permanently or temporarily exempted from the profit-sharing law.[34] The reforms, in addition, dealt with the mechanics of the National Profit-Sharing Commission that the 1961 constitutional amendment had empowered to decide the specifics of the new system. The duties of the commission, the qualifications of its members, and the manner of their selection were discussed. Finally, the procedure that workers were to follow should they wish to object to the amount of profits they were receiving was specified.[35]

An analysis of the reforms indicates that the president accepted some of the suggestions of both the private sector and organized labor. Both groups had suggested that the workers' share of profits be distributed approximately two months after the due date for income-tax declarations. This suggestion was accepted. The private sector had asked that new firms be exempted from the profit-sharing law for a number of years and that only the National Profit-Sharing Commission be given the power to revise the percentage of profits to be distributed. Organized labor had requested that collective contracts between employers and workers that would reduce the profit-sharing percentage established by the National Profit-Sharing Commission be null and void and that half the amount of profits distributed be based on the number of days worked and the other half be directly proportional to a worker's wages. These suggestions were accepted by the government.

The request on the part of both the business and labor leaders, however, that their respective representatives to the National Profit-Sharing Commission be selected by the main business and labor organizations was denied. Also denied was the private sector's suggestion that a worker's annual share of profits should not exceed 6 percent of his annual wages and organized labor's suggestion that disputes be reconciled by the Central Junta of

Conciliation and Arbitration, rather than by the Treasury.

Several demands of both parties were neither specifically accepted nor rejected by the government because they were not relevant to the reforms themselves. Many of the private sector's demands were in this category, because they related to the specifics of the future profit-sharing system, matters that were scheduled to be considered by the National Profit-Sharing Commission in 1963.[36]

The reforms of the Federal Labor Law were unanimously approved by both the Chamber of Deputies and the Senate in December 1962. Some of the provisions regarding profit sharing were, however, modified by the Chamber of Deputies. Directors of business enterprises were forbidden to receive a share of a firm's profits, as the original reforms had proposed. In addition, the time allowed for the elaboration of the specific profit-sharing system was reduced from two years to one year.[37] This last change caused some concern among government officials, who were not certain that the enormous amount of work involved could be completed in one year. The fact that the shortened time period would allow the finished product to appear during López Mateos's final year in office, however, probably caused him to have ambivalent feelings regarding the modification. As a result, it was accepted.[38]

The Reactive Role of Organized Interest Groups

Congressional approval of the Federal Labor Law reforms in December 1962 signified the end of the early phases of the profit-sharing decision. During this period, the leaders of the principal functional groups, organized business and labor, played an essentially reactive rather than initiating role vis-à-vis the government. Only in terms of their behavior toward the rank-and-file members of their organizations did the leaders take the initiative, principally by mobilizing their support.

Activities of the Private-Sector Organizations

The private-sector organizations at no time publicly criticized the principle of profit sharing, since it was a constitutional right.[39] They directed their efforts instead toward reducing the potentially detrimental effects that the future profit-sharing system might have upon the private sector. That the leaders of the private sector genuinely feared the impending profit-sharing system is clear from the language of a CONCAMIN memo that was drafted but never sent to the member chambers. The memo characterized the profit-sharing reforms as "laudable, just, Christian" but indicated that they "would bring industries to fathomless depths" [llevaría a las empresas a abismos insondables] because the proposed National Profit-Sharing Commission "will set the rules, which means that the executive will decide the percentage [of profits to be distributed]."[40]

The first order of priority was to assure that the government would take the views of the private sector into consideration, something it had not done during the preparation of the 1961 amendment to the Constitution. Soon after the 1961 reforms had been approved by Congress, therefore, the leaders of the principal private-sector organizations, in paid advertisements and in commentaries submitted to the press, began to criticize the manner in which the 1961 reforms had been prepared.

A full-page advertisement in *Excelsior*, which was paid for by COPARMEX, is representative of the tone and issues raised in these advertisements:

> COPARMEX profoundly laments the procedure followed [regarding the reforms to Article 123] of sending to Congress a constitutional amendment without previously soliciting the authorized opinion of the private sector, whose interests are affected by the above-mentioned project. We denounce the surprising and hasty procedure that has been followed to amend precepts of the Constitution that require a serious and conscientious technical study by our legislators, who, in any case, should have solicited the opinion of the affected sectors and of specialists in the material.[41]

The magazine *Siempre!* reported that "CONCAMIN complained it had not been consulted [regarding the 1961 reforms] and asked that its point of view be taken into account for the Federal Labor Law reforms."[42] And in the already cited unsent memo, CONCAMIN angrily accused the Chamber of Deputies and the Senate of acting in a "hasty" and "reprehensible" manner in approving constitutional reforms "without allowing time for the vital forces [*fuerzas vivas*] of the country to express their opinions.[43]

The second step in the strategy of the leaders of the private sector involved a careful study of the constitutional amendment, and the collection of data to support the argument that the profit-sharing system adopted should not be radically redistributive. The leaders of CONCAMIN took the initiative and called an emergency session to discuss the amendment, which was attended by the confederation's main officials.[44] It was decided that copies of the amendment were to be given to two of CONCAMIN's permanent commissions (Legislation and Labor and Social Security) for more detailed scrutiny.[45] The two commissions began to hold regular meetings in the middle of January in 1962 and the meetings continued throughout the year.

At approximately the same time, at the initiative of CON-CAMIN, the presidents of CONCAMIN, CONCANACO, CO-PARMEX, and the Asociación de Banqueros Mexicanos (ABM) held a series of private conversations. During these meetings they decided to form a "gran comisión" (great comission) or "comisión especial" (special commission) to be composed of several representatives from each of the four organizations.[46] CONCAMIN served as the coordinator,[47] and one of its officers, José Campillo Sainz, was the commission's president. CONCAMIN played this important role because its membership, which included all the major industrialists of Mexico, would be the group most affected by the profit-sharing system. Thus the CONCAMIN leaders had the greatest interest in convincing the government to accept the views of the private sector.

The members of the gran comisión were chosen by the four private-sector organizations. Each organization selected three to six representatives and was, in addition, represented in most cases by its president. Although some representatives were changed or added after the initial selection, the membership of the commission remained fairly stable. CONCAMIN's representatives, many of whom were also members of the two CONCAMIN commissions that were studying the constitutional amendment, were among the most important members of the gran comisión.[48]

The importance of CONCAMIN's representatives to the gran comisión is evident from the fact that five of them, José Campillo Sainz, Rafael Lebrija, Fernando Yllanes Ramos, Ricardo García Sainz, and Ramiro Alatorre, subsequently became members of the National Profit-Sharing Commission. In contrast, only two of CONCANACO's gran comisión members, Heriberto Vidales and Genaro García, eventually became members of the National Profit-Sharing Commission. The three remaining private-sector seats on the National Profit-Sharing Commission were filled by individuals who had not been members of the gran comisión.

It should be noted that all of the CONCAMIN representatives can also be considered representatives of COPARMEX, since the latter includes major industrialists among its members. The individuals in question, however, conceived of themselves more as representatives of CONCAMIN, the organization recognized by the government as the official spokesman of Mexican industrialists, than as representatives of COPARMEX, a voluntary association that was the unofficial spokesman of employers in general.

One of the first decisions that the members of the gran comisión made was to send their opinions regarding profit sharing to the state legislatures, which were then considering ratification of the 1961 constitutional amendment.[49] The views of the private sector were transmitted by the leaders of the local chambers of industry, chambers of commerce, and employers' centers located in each state. There was never, however, much hope that the state legislatures would heed the opinions of the private-sector organizations, since business leaders recognized that "when the president amends the Constitution, it is usually ratified by the state legislatures."[50]

Once the state legislatures had ratified the amendment, the private-sector organizations' alternate strategy of getting the government to "soften the consequences of the reforms by means of a serene and judicious [labor law reform]" went into effect.[51] A prominent labor lawyer and professor of labor law, Agustín Reyes Ponce, was commissioned in March 1962 by the members of the gran comisión to do a study of the profit-sharing reforms in which he was to "indicate the reasons why it was not practical, at the present time, to implement the [proposed] reforms . . . ." If the government failed to heed this opinion, the study was to elaborate a series of reform laws that would make the Federal Labor Law congruent with the newly amended Article 123 of the Constitution.[52]

The plans of the private-sector leaders had to be somewhat modified in May 1962 when the secretary of labor invited CONCAMIN and the other leaders of the private sector to present their views to the government commission that was in the process of drafting the reforms to the Federal Labor Law.[53] The deadline for the meeting between the private sector and government representatives originally was scheduled for June 15, 1962 and later extended to June 30, 1962 at the request of the business representatives.[54]

COPARMEX characterized the government's invitation as a response to the private sector's criticisms of the government's failure to consult its leaders in the drafting of the 1961 constitutional amendment,[55] although the government obviously made no substantial concession through such consultation. CONCAMIN's immediate response to the government's invitation was to call a meeting of the gran comisión, which had been meeting approximately twice a month. At a meeting in early June its members decided to discard their original idea of presenting a specific labor law reform project and to focus instead upon general points of concern to the private sector.[56] CONCAMIN's president emphasized that the private sector should send the secretary of labor only concrete suggestions regarding the regulation of the reforms. He considered it "unnecessary to combat the principle itself, since most of the state legislatures have approved the [1961 amendments to the Constitution]."[57] The study that had been commissioned in March was to be used as a working paper from which ideas were to be drawn.[58] Several additional meetings of the gran comisión were held. Finally, on July 2, 1962, representatives of CONCAMIN, CONCANACO, and COPARMEX met with the secretary of labor, informed him of the private sector's views regarding the manner in which the profit-sharing system should be enforced, and engaged in "an ample exchange of impressions" with him.[59]

In this meeting the private-sector representatives presented the secretary of labor with a written study that focused on the technicalities of the profit-sharing system to be adopted. It asked that a certain percentage of profits be excluded from profit sharing in order that investors receive a fair return on their capital investment. The study also suggested that federal, state, and municipal taxes (including income taxes) be deducted and the remaining money be considered a company's profits. It recommended that a worker's share of profits should not exceed 6 percent of his annual wages and suggested that new firms should be exempt from profit sharing during the first five years of their existence. Other opinions dealt with the manner of selecting representatives to the 1963 National Profit-Sharing Commission and the way profits were to be distributed. As has been noted, some of these suggestions were incorporated into the Federal Labor Law reforms; others were rejected; and the rest were not relevant to the subject under consideration.[60] Many of the activities of the private sector during this period were not focused solely upon the profit-sharing issue. The 1961 constitutional amendment was also concerned with such issues as minimum wages and *reinstalación forzosa*, the obligatory reinstatement of workers who have been dismissed by their employers without just cause. Reinstalación forzosa and profit sharing were the two issues of greatest concern to the private-sector organizations; the gran comisión and CONCAMIN's two permanent commissions on Legislation and Labor and Social Security dealt with both issues. After the private sector had presented its views on profit sharing to the government in early July, therefore, the gran comisión completed its study of obligatory reinstatement of workers. Its views on this matter were presented in mid-July during another meeting between the secretary of labor and the representatives of CONCAMIN, CONCANACO, COPARMEX, and the ABM.[61] The private sector's final study, which concerned minimum wages, was given to the secretary of labor by representatives of CONCAMIN, CONCANACO, and COPARMEX at the end of August.[62]

The members of the private sector organizations had not requested their leaders to take action. The leaders had decided to act and had then informed their members of their activities in an attempt to mobilize them to support the leaders' efforts to make the future profit-sharing law more palatable to the private sector. The extent to which the leaders attempted to mobilize their followers varied according to the leaders' perception of the interest they and their members had in the matter.

Because CONCAMIN's membership consisted exclusively of industrialists and included the most important ones in Mexico, CONCAMIN's leaders undertook the most intensive mobilization effort. The CONCAMIN leaders sent circulars describing the private sector's activities concerning the issue to their member groups at frequent intervals throughout 1962, and the organization's biweekly publication, the *Boletín Quincenal*, included some of the same information. In addition, profit-sharing studies done by other individuals and organizations were occasionally disseminated to the member groups. COPARMEX, which included among its members many of the more important members of CONCAMIN, also undertook a fairly intensive mobilization effort. In October 1962 it held a round table discussion during which the views that the private sector had expressed to the government regarding profit sharing were restated.[63] The organization's public relations department issued periodic bulletins that told of the efforts of the private sector to prepare studies and to make its views known to the president and the secretary of labor. The COPARMEX and CONCAMIN circulars also sought to make their members share the views of their leaders; subjective opinions were liberally sprinkled among the more objective reports of private-sector activities. CONCANACO, whose members were individuals involved in commercial activities, showed little interest in mobilizing its rank and file. Only one circular dealt with the activities of the private sector, and the fact that CONCANACO was participating in the studies of the gran comisión was noted briefly in a publication regarding the Forty-fifth Regular General Assembly of CONCANACO. The ABM informed its members of its efforts in its main publication, the *Revista Bancaria*, but it provided few details.

When President López Mateos sent the proposed reforms to the Federal Labor Law to Congress in December 1962, the reaction of the private sector was far different from the criticism with which it had responded to the president's 1961 constitutional amendment. The president of COPARMEX, for example, praised "the constructive collaboration that existed among the government, workers, and businessmen for the formulation of the Federal Labor Law reforms." He further stated that the government's 1962 reforms had "incorporated the majority of the points of view of COPARMEX and the other private-sector organizations."[64] The private sector thus was satisfied that its efforts had not been in vain. The government, by incorporating the business leaders into the 1962 phase of the decision-making process, had achieved its goals of repairing its relations with the private sector.

The Response of Organized Labor

Mexican labor leaders reacted to the 1961 constitutional amendment by strongly praising both the amendment and President López Mateos. The leaders used various forums to express their approval—the newspapers, labor congresses, parades and rallies, personal meetings with the president that were well reported by the press, and conferences. Even labor leaders who had never demanded profit sharing and who, in fact, had harbored strong reservations concerning it hailed the reforms as "the just answer of a labor-oriented president to the petitions and demands that the working class has presented for many years."[65]

The unreserved praise was supplemented throughout 1962 by efforts to mobilize rank-and-file members in support of the reforms. In their mobilization efforts the leaders prepared articles for the newspapers or publications of the different labor groups and organized conferences that were "intended to diffuse among the workers and people interested in labor law the true significance and transcendence of the reforms to Article 123."[66] The aspect of the reforms that needed most explanation concerned the relationship between profit sharing and the class struggle.[67]

The leaders of organized labor also attempted to mobilize the support of their rank-and-file members by portraying the private sector as a staunch opponent of profit sharing and the reforms in general and as an enemy engaged in surreptitious efforts to sabotage them. The rationale behind the strategy was that the rank and file would rally around its leaders and the reforms more willingly if the workers perceived that their interests were threatened as a result of concerted attacks by a powerful and united private sector. The labor leaders' task, however, was made more difficult by the fact that the leaders of CONCAMIN, CONCA-NACO and COPARMEX had taken great care to avoid publicly attacking the *principle* of profit sharing.[68] To establish the validity of the threat, therefore, the labor leaders publicized the private sector's early efforts to convince the state legislatures to reject the amendment as well as certain publicly stated views (for example, the request for a maximum limit to the amount of profits a worker could receive) as "proof" that the private sector was opposed to profit sharing. The highly critical attitude of extremist private-sector groups like the Employers Center of the Federal District, which did not belong to CONCAMIN, CONCANACO, or COPARMEX, was given a great deal of publicity by labor leaders, who obviously made no effort to explain to their rank and file that such criticism could not be considered representative of the principal private-sector organizations.[69] Rumors that the private sector intended to ask the government to schedule collective contract revisions every five years (the prevailing system required revision every two years) because of the impending establishment of a profit-sharing system were also exploited by the labor leaders in order to mobilize their followers.[70] Finally, the private sector's vociferous objections to a specific section of the amendment that provided for the obligatory reinstatement of unfairly dismissed workers was portrayed by the labor leaders as an attack on the entire constitutional amendment.

Although the leaders of organized labor deliberately cultivated the specter of a united and powerful private sector determined to sabotage the prolabor reforms, they apparently did not themselves perceive the threat to be severe. If they had, they would have forgotten political rivalries and joined forces against the private sector, as they eventually did in 1963. In 1962, however, they refused to cooperate. As a result, each of the two main organizations into which the Mexican labor movement was divided, the BUO, headed by the CTM, and the CNT, led by the CROC, held separate study sessions in mid-November. The resulting opinions were formally presented by each group to the secretary of labor the same month.[71] The purpose of these study sessions was supposedly to elaborate the points of view of the BUO and the CNT. However, the BUO and CNT leaders had already presented their views to the secretary of labor several weeks before,[72] and the sessions were clearly only another device to mobilize the rank-and-file laborers.

The Peripheral Role of the PRI

The PRI cooperated with the leaders of the organized labor movement in their efforts to mobilize support for the president's labor reforms. Party leaders joined with the labor leaders in criticizing the private sector's efforts to make the reforms more palatable to business interests. The PRI's official publication, *La República*, and the main newspapers, especially *El Nacional*, the government newspaper, were the vehicles used by the party's leaders to express their views and to applaud the president's decision.[73] This limited supportive role was the only one the official party played during any phase of the profit-sharing decision.[74] There is thus no evidence in this case to support the assertion of one early analyst of the Mexican political system that "the effective decision-making takes place . . . in the interaction of interests in and around the revolutionary party . . ."[75] In the case of the profit-sharing decision, the interest groups belonging to the PRI (i.e., the labor groups) did not work through the labor sector of the party. Instead the labor leaders completely bypassed the party and dealt directly with the secretary of labor. The chambers of commerce and industry were not involved in any way with the PRI. They worked with the secretary of labor. Although organized labor officially belongs to the PRI and the private-sector organizations are officially excluded from membership, these facts have no importance with regard to the profit-sharing decision.

The Politics of the National Profit-Sharing Commission

5

The 1961 amendment to Article 123 of the Constitution had stipulated that the obligatory profit-sharing system to be adopted was to be elaborated by a National Profit-Sharing Commission composed of representatives of the government, the private sector, and organized labor. The commission was to be an independent body and would cease to exist upon issuing its decision regarding the specifics of the profit-sharing system. The commission's decision, according to the 1962 Federal Labor Law reforms, was to be reached by December 1963, and it was to be based on the commission's studies and investigations of the Mexican economy. By the end of December 1962, President López Mateos was in a position to form the National Profit-Sharing Commission, for both the Constitution and the Federal Labor Law had been duly amended.

The Formation of the Commission

On February 4, 1963, the secretary of labor announced the appointment of Hugo B. Margáin as president of the National Profit-Sharing Commission. Margáin, a former head of the Income

94

Tax Department of the Treasury, was serving as undersecretary of the Ministry of Industry and Commerce.[1] The executive's selection of an apolitical technocrat as head of the commission served to emphasize the fact that the remaining decisions concerning the profit-sharing system, such as the amount of profits to be distributed and the amount of return to be allowed on a capital investment, were technical ones that were to be made on technical and not political grounds.

Once Margáin was appointed, President López Mateos delegated decision-making power to him. Margáin thus would have complete discretion in deciding the way in which the government would vote on the tripartite commission. This delegation of decision-making power by the Mexican president was not only specific (i.e., it only involved profit sharing), but also temporary, for the president would reclaim it when the commission completed its work. It was also understood that despite this delegation of decision-making power, the head of the commission, as a trusted legal subordinate of the Mexican president, was always to defer to the president should a difference of opinion arise between them.[2]

Despite the restrictions inherent in the delegation of decision-making power, the leeway that Margáin possessed during this stage of the profit-sharing decision was considerable. He enjoyed complete discretion in the selection of his technical staff. Furthermore, the president apparently made no reccommendations or suggestions regarding the kind of profit-sharing system he wished established. His principal concern had been with the political aspects of the decision (the "decision in principle"). He gave little thought to the technical (mainly economic) problems that his decision would create.[3] Raymond Vernon has noted that "when the instructions from above are ambiguous or when the situation calls for technical action in the absence of instructions, the [considerable] power of the [Mexican] technician is enhanced even further."[4] Therefore, the decision of Margáin and his technical staff would have a decisive impact on the type of profit-sharing system eventually adopted in Mexico.

Most of the individuals whom Margáin named to his technical staff were legal and fiscal experts. All of them had worked or were still working in a government ministry in a technical capacity. Margáin selected them as members of the technical staff of the National Profit-Sharing Commission in part because of their

technical qualifications. A more relevant and important consideration, however, was the fact that they were all loyal and trusted
subordinates of Margáin or of his old friend Octavio Hernández,
whom Margáin had appointed to the second-ranked position on the
technical staff. The Margáin appointees were thus consistently
referred to collectively, both by each other and by third parties, as
"el equipo de Margáin" (Margáin's team). Had another man been
chosen to head the commission, a different team of experts loyal to
that individual would have been selected.

At approximately the same time that the president selected
Margáin to head the National Profit-Sharing Commission, the
secretary of labor invited the members of the private sector and the
labor movement to participate in separate elections in which each
group would choose five representatives and five alternates to the
commission. The labor unions or businesses were to select delegates
who were to vote only in the election which corresponded to their
branch of industry.[5] Each labor or business delegate would have as
many votes as the number of workers or businessmen he
represented.

When they learned of the impending elections, the leaders of the
private-sector organizations continued their cooperative behavior.
They decided to present a single list of candidates, with the proviso
that each candidate had to be acceptable to all of the private-sector
leaders involved. The leaders of the various private-sector organizations then either selected the delegates, presented their rank-and-
file membership with a list of potential delegates from which the
rank and file were to select one, or suggested that each chamber of
a particular confederation authorize its head to act as a delegate,
with the parent confederation reserving the right to substitute two
members of its executive commission for the delegates, "should it
prove necessary to do so."[6] The delegates were then instructed to
vote in favor of the single list of candidates that would be presented
jointly by CONCAMIN, CONCANACO, and COPARMEX.[7]

The leaders of CONCAMIN, CONCANACO, and COPARMEX
claimed their presentation of only one list of candidates was
justified because the private sector was united on the profit-sharing
issue and therefore could and would speak with one voice. They
further stated that as a result of this unity the ten private-sector
members of the commission (i.e., the five representatives and the
five alternates) were not representing the specific organizations to
which they belonged (i.e., CONCAMIN or CONCANACO) but

the entire Mexican private sector. Nevertheless, an examination of the representatives' affiliations indicates that CONCAMIN exercised the greatest influence on the private-sector delegation, for it was unofficially represented by six individuals: José Campillo Sainz, Carlos Isoard, Ramiro Alatorre, Ricardo García Sainz, Rafael Lebrija, and Fernando Yllanes Ramos. Some of these men were also members of COPARMEX, but considered their primary allegiance to lie with CONCAMIN. One representative, César Roel, was specifically identified with COPARMEX. The remaining three representatives, Heriberto Vidales, Genaro García, and Alfonso Ortega Vélez, were members of CONCANACO.[8] The private-sector leaders had complete autonomy in their choice of representatives. The government's only suggestion was that businessmen be given the regular seats on the commission and lawyers be made the alternates.[9] The suggestion was made in the interests of harmony and perhaps justice, for the labor leaders might have objected to negotiations in which lawyers rather than businessmen were their adversaries. The private sector accepted the government's suggestion. This concession did not mean, however, that none of the regular private-sector representatives were lawyers, for several of the businessmen also held law degrees. Furthermore, the distinction between regular and alternate representatives was somewhat illusory, since at the commission's first meeting in March 1963, it was formally decided that no distinction would be made between the two types of representatives. All ten could speak and vote on the commission. This rule also applied to the ten labor representatives.

The businessmen who represented the private sector managed several of Mexico's biggest and most important firms.[10] This bias was not surprising, since most of the businessmen were affiliated with CONCAMIN, which is dominated by the largest firms, whose heavy financial contributions entitle them to greater voting power. The five alternates were well-known labor lawyers who represented some of the largest private-sector firms in Mexico. Most of the ten representatives were very active in either CONCAMIN, CONCANACO, or COPARMEX and frequently served on their executive commissions. Several of them had been president or vice-president, or both president and vice-president, of one of the three organizations. The talents and experience of the ten representatives were varied, but relevant to the task they were undertaking. Some were professors of tax or labor law and one was an accountant. Two of the representatives, prior to entering the private sector, had worked in the Treasury's Income Tax Department under Margáin, and five had played active roles in CONCAMIN's two commissions of Legislation and Labor and Social Security during 1962, when the commissions were making a detailed study of the profit-sharing issue.

In contrast to the unified response of the leaders of the private sector, the announcement of the impending elections for representatives to the National Profit-Sharing Commission heightened the existing rivalry among the leaders of the organized labor movement. The leaders of the CTM antagonized all the others by claiming that they could win all of the labor seats on the commission because they were the largest labor confederation in Mexico.[11] The CTM leaders were assuming that the other labor leaders would be unable to forget their differences in order to unite and defeat the CTM candidates.

At least one of the leaders of a non-CTM labor group expressed to the government his group's dissatisfaction over the method of selecting labor's representatives to the commission. He suggested that a way be found to assure representation for non-CTM groups. The government reacted favorably to the suggestion, and a government spokesman spoke with the CTM leaders. Shortly afterward, the CTM agreed to allow other labor groups to be represented on the commission.[12] The CTM leaders explained that they had decided to allow minority representation because they wanted to assure that the minority groups "do not alienate themselves from the problems of the working class as they have on other occasions." The problems and interests of all workers' organizations were the same, the CTM leaders added. Furthermore, giving seats to the minority groups would promote "a spirit of solidarity."[13] They also offered the additional explanation that the minority groups, "which have poorly understood the transcendence of the constitutional amendment, should share responsibility for the studies and the decision regarding the [percentage of profits to be shared]."[14]

The last explanation is probably the most valid. The government had convinced the CTM leaders that labor unity was necessary in order for the government to assure that the workers' right to profit sharing was not destroyed by a united and mobilized private sector actively seeking to protect its own interests. Such labor unity was impossible unless all labor groups were made to feel that they had a stake in the profit-sharing system as a result of their participation in the deliberations of the National Profit-Sharing Commission.

The CTM's willingness to allow minority representation on the commission decreased antagonism toward it and brought praise from labor groups that until then had rarely had anything good to say about the CTM. The left-wing FOR, for example, usually a severe critic, declared "that such positive and encouraging gestures toward unity . . . merit not only applause and support, but should also be reinforced and repeated throughout the country."[15]

Following the CTM decision, the leaders of the BUO and the CNT, the two large blocs into which the organized labor movement was divided, met and decided which groups were to be represented on the comission. It was agreed that the BUO, which was dominated by the CTM and included a majority of Mexico's unionized workers, was to receive a majority of the seats. A single list of candidates acceptable to all of the labor leaders was then prepared and presented to the secretary of labor, who had indicated several labor leaders whose inclusion would be appreciated by the government.[16] The various labor groups then staged elections during which the persons whom the labor leaders had agreed would represent each labor organization were formally elected by their respective organizations. A short time later, the government-sponsored election to select the ten labor representatives was held, and the list of candidates that had been privately agreed upon by the labor leaders was the only one presented. All of the individuals whose names appeared on this list were unanimously elected by the 4,682 delegates, who represented 114,523 workers.[17]

The labor representatives were Adolfo Flores Chapa and Alfredo Rodríguez of the Miners' Union; Jesús Yurén Aguilar and Blas Chumacero of the CTM; Enrique Rangel and Alberto Juárez Blancas of the CROC; José Ortiz Petriccioli of the CROM; Samuel Ruiz Mora of the FOR; Francisco Ballina Tabares of the Pilots' Association; and Francisco Benítez of the Theatrical Federation. The majority of the labor organizations represented were affiliated with the BUO. The exceptions were the CROC and the FOR, which belonged to the CNT. The only important national labor unions that did not elect representatives to the commission were the Electricians and the Railroad Workers. Their absence was probably the result of the fact that both the railroads and the electric-power industry were government-owned. It would not have been considered legitimate for the government to have union representatives from government-owned industries representing the labor movement on a tripartite commission on which the government would cast the deciding vote.

The individuals chosen to represent organized labor on the commission were, in general, older men in their fifties and sixties who had been active in their respective confederations since their creation. Some had been leaders of the Mexican labor movement since the early 1930s. These men were well known both within the labor movement and by high-ranking persons in the government and in the private sector. Their names were often considered synonymous with the labor organizations they represented.

The Role of the Técnicos

The Federal Labor Law reforms of December 1962 had stipulated that the National Profit-Sharing Commission was to hold its first meeting on March 1, 1963. This stipulation presented a serious problem for the comission's president and his technical staff because almost none of the technical information that they required in order to decide upon a specific profit-sharing system was readily available. The three-person commission that had prepared the 1961 constitutional amendment and 1962 Federal Labor Law reforms had been concerned with more general and basically historical and legal questions. It had not had to undertake the extensive economic studies required in order to make a responsible decision regarding the technical aspects of a profit-sharing system. Thus, one of the first tasks undertaken by the technical staff was the collection of technical and statistical data from the Mexican Treasury, especially its Income Tax Department, the Ministry of Industry and Commerce, and the Presidency.[18]

Despite the delegation of decision-making power by the president to Margáin and his technical staff, the técnicos originally conceived of their role as limited to the collection of data. Presumably this data would then be used to reconcile differences between the profit-sharing projects proposed by the labor and business representatives. The técnicos did not at first have any intention of elaborating their own profit-sharing system. It became evident very early, however, that specific profit-sharing projects might not be forthcoming from either the private-sector representatives or the representatives of organized labor. As a result, the technical staff realized that not only would it have to collect basic data but it would also have to propose an original profit-sharing system that would eventually have to be approved by both groups of representatives.[19]

The members of the technical staff saw their newly defined task as two-pronged: to propose a viable system that would enable the workers to share in profits and to construct a system that would give added impetus to economic growth, or at least avoid slowing it. Although López Mateos primarily had focused on the political implications of profit sharing, the technocrats were most concerned with the economic implications.[20] The profit-sharing decision thus had become somewhat modified as it moved from its first to last phases. What had originally been a political decision for the benefit of organized labor had subsequently also become a decision whose purpose was to give impetus to economic growth.

The Mexican preoccupation with economic growth, which was equated with industrialization at least until the early 1960s, started when "the impact of World War II began [to be] felt, giving Mexico a major opportunity to begin realizing its industrial possibilities."[21] In 1940 "the first of a succession of presidents devoted to the proposition that industrial growth on the modern pattern was indispensible to Mexico" was elected.[22]

The new concern with economic growth or industrialization was reflected in *El Trimestre Económico*, a leading economic journal that enjoys wide distribution throughout Latin America. In the late 1940s, the journal began to publish numerous articles by prominent Mexican and Latin American economists discussing the need for industrialization and the methods for achieving it. The most celebrated and definitive statement of the need to industrialize and the manner to achieve industrialization appeared in the journal in 1949. The author, Raul Prebisch, basing his recommendations on the fundamental assumption that the terms of trade for exporters of raw materials (e.g., Latin America) were continually deteriorating, stated that industrialization could be achieved through an enormous accumulation of capital, well-directed foreign investment, and import substitution. This strategy would increase the need for imported capital goods, which, in turn, would necessitate savings and a reduction in consumption.[23]

The "Prebisch thesis" gained wide acceptance among Latin America's nationalistic intellectuals, businessmen, and técnicos. Mexico's técnicos, although products of the Mexican Revolution, basically subscribed to the Prebisch thesis and its derived prescriptions. They wholeheartedly believed that Mexico had to industrialize. If it failed to do so, they felt, it would continue to be a victim of deteriorating terms of trade. In order to industrialize, capital investment, preferably domestic rather than foreign, was considered a necessity.[24] The main preoccupation of the members of the technical staff of the National Profit-Sharing Commission, therefore, was to elaborate a profit-sharing system that would enable workers to share in profits but that would also not prove detrimental to the investment process. As the head of the commission later wrote, "What was the main purpose [propósito medular] of the reform? The essential thing in the economic sphere is the investment process. Everything else is derived from it."[25]

Before work on the specifics of a profit-sharing system could begin, therefore, there had to be agreement on what percentage of the private sector's profits could be distributed to the workers without decreasing that sector's investment in the Mexican economy. Margáin and the members of his staff, after discussing the problem with high-ranking officials in the Treasury, the Ministry of Industry and Commerce, the Bank of Mexico, and Nacional Financiera (the national development bank), decided that 5 to 7 percent of total private profits could be distributed without disturbing the investment process. This percentage was the equivalent of 500 to 700 million pesos based on the private sector's profits for 1961-1962, which, according to the government, equaled approximately 10 billion pesos. The percentage was discussed with the private sector's technical advisers and commission representatives, who found it acceptable.[26]

The next step was to determine the amount of profits made by individual firms operating in Mexico. The Treasury made all 600,000 income-tax returns available to Margáin and his staff, who did a computer analysis of 14,000 of them. The sample included firms of all types, located throughout the country. Once the general information had been obtained and studied, numerous formulas for distributing profits to the workers were tested in order to discover the amount of additional income workers employed by each of the 14,000 firms would receive in each case.[27] One formula was eventually adopted. It allowed business three deductions, based on the amount of annual total profits rather than on the total capital investment of a firm. The first deduction represented the *interés razonable* on capital investment to which the private sector was entitled according to the 1961 profit-sharing amendment. The second deduction was allowed for funds used for reinvestment. The third deduction varied according to the relation between the total capital investment of a firm and the total amount of money paid to its labor force. The greater the ratio of capital to labor, the larger the allowable deduction. The most mechanized, capital-intensive firms thus would be entitled to the greatest deduction. This third deduction had two purposes. The first was to ensure that workers in large, modern and highly profitable industries would not receive a great deal of money while their counterparts in older, more labor-intensive industries received very little money. The second purpose was to give impetus to the increased modernization of Mexican industry, since by becoming more capital-intensive, a firm would have to share a smaller percentage of its profits. Once the three deductions were made, all workers would be entitled to receive the same percentage of the remaining profits. On December 11, 1963, Margáin presented this profit-sharing proposal to the labor and business representatives on the National Profit-Sharing Commission.

The Meetings of the Commission

At the same time that Margáin organized and began working with his technical staff, he presided over the meetings of the National Profit-Sharing Commission, which was formally inaugurated on February 28, 1963. The ostensible function of the commission was the elaboration of a profit-sharing system for Mexico. Its real function, however, was somewhat different.

President López Mateos had no real technical reason for establishing such a commission. He could have charged one of the government ministries with the task, incorporated the system it proposed into his 1962 reforms to the Federal Labor Law, and sent the legislation to Congress for its approval. That procedure, however, would have done nothing to demobilize the opponents of the profit-sharing decision or to mobilize support for it. In fact, such behavior on the part of the Mexican president might have increased opposition and polarized the groups that believed themselves adversely affected by the decision (the private sector groups), as well as the theoretical beneficiaries of the decision (the members of the organized labor movement).

The formation of the National Profit-Sharing Commission performed a support-building function. As a result of their participation on the commission, the private-sector leaders might ultimately come to support a profit-sharing system for which they were partially responsible and they might then encourage their followers to support it also. From the government's point of view, it was also necessary to mobilize support among the leaders of the organized labor movement, some of whom were less than enthusiastic about the decision. Increased support by the labor leaders would facilitate their efforts to mobilize the support of their rank-and-file members.

The National Profit-Sharing Commission also served to unite the beneficiaries of the profit-sharing decision. The private-sector leaders had joined together to protect their interests immediately upon learning of the president's decision to establish an obligatory profit-sharing system. The labor leaders, however, had remained divided among themselves. They saw little reason to reconcile their differences because they were the theoretical beneficiaries of the president's decision. Because the decision was in a sense a gift to them from President López Mateos, a gift that they had not worked specifically to obtain, they assumed that the president would protect the interests of the workers against challenges to his decision from the private-sector leaders. President López Mateos, on the other hand, would be better able to withstand the attacks of the private-sector leaders if he could confront them with the specter of a labor movement strongly united and highly mobilized in support of profit sharing. The National Profit-Sharing Commission was thus a device intended to force a confrontation between the divided labor leaders and the united private-sector leaders to cause the labor leaders to forget their differences and unite.

The establishment of the National Profit-Sharing Commission had another political justification. It would reconcile the two groups of representatives with each other. For the future profit-sharing system to function smoothly, it would be necessary to diminish the antagonism between the private-sector leaders and those of the organized labor movement, which had intensified during 1962.[28] Such reconciliation was also necessary for the continued viability of the Mexican regime.

Under the leadership of Margáin, the commission succeeded in attaining its informal goals. From the very beginning, Margáin attempted to undermine the belief held by the business representatives on the commission that the profit-sharing reform was "antibusiness." Instead, he sought to convince both business and labor leaders that profit sharing would benefit both groups and ultimately the Mexican nation. At the second meeting of the commission, he stated categorically that there was no place for a "class struggle" attitude on the part of the commission representatives. Workers and businessmen, he said, should work with the government in the pursuit of a single goal—the formulation "of a resolution that meets the needs of the country and that fulfills another of the social justice goals of the 1917 Constitution." In order to attain this goal, Margáin continued, "our only interest is the economic development and social justice of Mexico, which includes all of us and supersedes any concrete interest." He concluded by citing "the broad field of similar interests among the representatives that exists whenever matters of national interest are under consideration."[29]

Margáin's attempts to create a national perspective among the commission's members had a motivation similar to the executive's effort to involve as many groups and interests as possible during this later stage of the profit-sharing decision. The involvement, however, was limited to the provision of information and points of view, and the government made no pledge to use the information or to accept the points of view. On March 1, 1963, the government published an announcement in the *Diario Oficial* inviting all interested parties to aid the National Profit-Sharing Commission by presenting studies for its consideration.[30] In July 1963, the law faculty of the National Autonomous University of Mexico, apparently at the suggestion of the government, sponsored a series of conferences on different aspects of profit sharing. Two of the four speakers were Juan Landerreche Obregón, a lawyer active in the PAN and Vicente Lombardo Toledano, the head of the Partido Popular Socialista (PPS). The Academy of Labor Law devoted its Fourth National Assembly in September 1963 to the question of profit sharing, and the College of Accountants of Guadalajara followed suit at its Fourth National Convention of Accountants in November 1963.

The government also used the media in its campaign to make profit sharing a national rather than a partisan issue and thereby mobilize support for it. Government officials and persons working with the government on the profit-sharing decision gave speeches and wrote newspaper articles stressing the progressive nature of profit sharing and the benefits it would bring to all groups and to the country in general. In July 1963, for example, an adviser to the commission's technical staff, in a speech before the American Chamber of Commerce in Mexico, told his listeners: "Mexico is going to add to her role of active supporter of peace in international politics the role of leader of international economic politics when profit sharing begins to operate both legally and officially."[31]

Once the tone of national unity had been set and the study committees had been organized, the meetings of the commission were devoted to the presentation of papers dealing with the ten themes into which the work of the commission had been divided. Most of these early papers were given by members of the technical staff, and the few that were not had been commissioned by the staff. The papers were mainly concerned with the Mexican economic situation and the historical antecedents of profit sharing in Mexico. Their purpose was to "educate" the labor and the private-sector representatives concerning the complexities of the profit-sharing issue in the hope that increased cooperation and support would result.

By the end of July, which represented the approximate halfway point in the commission's existence, the most important questions had not yet been discussed. Furthermore, the representatives of the private sector and organized labor had both submitted a written opinion or statement of their points of view on only four of the ten themes. Six of the themes had not been treated at all by any of the commission's three groups, labor, business, or the government.[32]

In October 1963, the commission finally began to focus upon the most problematical and crucial questions. The first major issue considered was the interés razonable—the rate of return on capital investment to which a businessman would be entitled. The commission devoted more time to this problem than to any other.

The government's position regarding the interés razonable had been hinted at in a July meeting of the comisión. At that time Margáin had expressed the view that a businessman's right to receive a reasonable return on his investment did not entitle him to deduct from his profits a certain percentage of the total capital investment of his enterprise. Margáin stated his opinion that such a deduction was unconstitutional because it could nullify the profit-sharing law. In many cases, he explained, a deduction of a percentage of a firm's total capital investment from profits would either exceed total profits or would leave very little of a firm's total profits for distribution among the workers.[33] Rather than allow a deduction based on a firm's total capital investment, Margáin favored allowing a deduction from total profits of a percentage of a firm's total profits. This system would ensure that some percentage of a firm's total profits would always be available for distribution among the workers.

The private sector's view regarding interés razonable was diametcially opposed to that of Margáin. At an October meeting private-sector representatives affirmed business's right to a "prior deduction," that is, a deduction from profits of a certain percentage of the total capital investment of a firm before applying the profit-sharing formula.[34] The private sector justified its demand in terms of the "interest of Mexico." "In the hierarchy of values in the Explanation of Motives of [President López Mateos's profit-sharing] intitiative was the necessity of fomenting the industrial development of the country," stated a representative of the private sector. "More important than anything, Mexico comes first, before anything else."[35] The private sector's attempt to make a partisan demand seem a national one was reminiscent of the government's earlier efforts to make its profit-sharing decision appear less in the interest of labor (and to the disadvantage of business) and more in the interest of the entire nation.

The private-sector representatives initially demanded a deduction equivalent to 12 percent of the total capital investment of an enterprise as a reasonable return on investment. When this proposal was rejected, the private sector reduced its demand to 9 percent. Margáin, representing the government, remained firm in his belief that a prior deduction based on total capital investment would be unconstitutional and would make the profit-sharing law meaningless. Margáin's resolution caused the private-sector representatives to reduce their demand to 6 percent and finally to 4 percent of the total capital invested, but Margáin did not alter his position.[36]

The private-sector leaders justified their demand for a 12 percent return on their investment by reminding Margáin that failure to allow that percentage would make it unprofitable to invest in industry. Other forms of investment would be more attractive because an equivalent or greater return could be obtained by purchasing certain kind of bonds or by banking one's money. They reminded Margáin that Mexican banks received 12 percent interest on their loans to the public and that Nacional Financiera received approximately 12 percent interest on its *títulos financieros* (a kind of bond). The lowest return on an investment at that time in Mexico was 4.5 percent, which was the amount of interest paid on short-term deposits with savings banks. The interest rate was 6.5 to 8 percent on longer-term deposits, 8 percent on financial bonds and mortgages, and approximately 9 percent on private loans. The private-sector leaders therefore considered their subsequent demands for a 9 percent or 6 percent return on their investment perfectly reasonable, and their final demand of 4 percent extremely generous. Margáin, however, stated that their argument was faulty, because if they were to put their money in Mexican banks rather than investing it in industry, there would be abundant capital in Mexico for investment and the interest rates would decline.[37]

The position of the labor representatives to the commission was the same as that of the government, represented by Margáin. In October, one of the labor representatives declared that although labor supported business's right to a reasonable return on its investment, a prior deduction from profits based on the amount of capital invested in a firm was unacceptable. Instead, the labor leaders supported Margáin's idea of a deduction of a certain percentage of the total profits as a fair return on the capital invested.[38] Labor stressed, however, that its position represented a concession in the interests of the nation, because labor would really prefer not to allow business any deductions from total profits.[39]

This "concession in the interests of the nation" was the result of private discussions between the labor representatives and Margáin. Margáin had expressed the opinion that a profit-sharing system that would not allow some kind of deduction from total profits would ultimately prove detrimental to the Mexican economy and, by extension, to organized labor. He also stressed that profit sharing was a new system that might possibly prove unsuccessful. It made little sense, therefore, for labor to make extreme demands at this early stage, because such behavior might destroy the system. The decision regarding the percentage of profits that labor was to receive could always be changed at a later date. It was therefore essential to ask for less now in order to assure the system's success. Once the system proved viable, labor could demand a greater share of profits.[40] In contrast to the representatives of the private sector who remained adamant in their demand for a prior deduction, the labor representatives proved less recalcitrant in their opposition to any kind of deduction and more willing to accept the reasoning of Margáin and his staff.

It is not surprising that the labor representatives were more acquiescent than their private-sector counterparts. As noted earlier, the Mexican labor movement is characterized by limited political pluralism. Its leaders serve two masters, their rank-and-file members and the Mexican president and his high-ranking subordinates. In the specific case of profit sharing, the labor movement was in a situation that optimized the president's opportunities to impose his will upon its leaders and their rank and file.

First, labor had received a "gift" from the president. The president had decided to establish a profit-sharing system "for the workers" despite the fact that the labor movement had not worked hard to obtain this benefit. Labor's representatives thus would have found it difficult to make extreme demands regarding the amount of profits the workers were to receive, even if they had been less dependent on the regime.

Second, the profit-sharing issue was complex, and a certain amount of technical training or knowledge was required if a person was to understand the economic and fiscal problems involved. Few of the labor representatives on the National Profit-Sharing Commission possessed such expertise, and the labor movement did not have the economic resources to hire many technically trained experts to determine its interests and help safeguard them even if it had wished to do so.

As a result, the representatives of organized labor turned to Margáin for guidance and support. The fact that the government was the source of the profit-sharing decision made the labor leaders think of Margáin, the president's representative, as their ally and friend. Their respect for Margáin seemed to reinforce their inclination to trust the government to protect their interests. Their tendency to trust him no doubt provided Margáin with increased leeway in his pursuit of the government's goals.

The representatives of the private sector, because they enjoyed greater autonomy than their labor counterparts, remained adamant in their insistence that business should be allowed to deduct a certain percentage of its capital investment from profits before the profit-sharing formula was applied. As a result, Margáin devoted a major portion of his energies to persuading the private-sector representatives to modify their position and to adopt the point of view of the government, represented by the technical staff and Margáin. Margáin's strategy involved taking a hard-line attitude regarding business's insistence on a prior deduction based on capital investment. Margáin categorically stated that the demand was unconstitutional and detrimental, and he left no doubt that the government would not consider it.[41] He then presented the plan of the technical staff, which, in essence, provided that whenever profits existed, a percentage of them would be distributed to the workers. To achieve this goal, any deductions for the purpose of granting a reasonable return on capital investment would have to be based on profits, rather than on the capital investment. Margáin thus presented his plan as a compromise between that favored by labor (no deductions from profits should be allowed) and that proposed by the private sector (a deduction from profits equalling a percentage of the total capital investment of a firm should be allowed, and the workers should receive a share of any profits that remained).

Another aspect of Margáin's strategy was his rejection of all attempts to substitute some other benefit for profit sharing. The private-sector representatives, for example, faced with Margáin's intransigence, had suggested that they would be amenable to distributing the equivalent of one month's salary to each worker in lieu of sharing profits. Several labor leaders thought this compromise was acceptable. Margáin, however, criticized it because it blurred the distinction between profits and salaries, and pointed out that one of the reasons obligatory profit-sharing systems had failed in other countries was an inability to make this distinction.[42] He spoke with the president and convinced him that the system he supported was better than the proposed substitute. López Mateos gave Margáin his full support,[43] and Margáin then reaffirmed his opposition. He also spoke privately with the representatives of organized labor and persuaded them that it was not in their interest to accept a month's salary for each worker as a substitute for profit sharing.

The most important aspect of Margáin's strategy to persuade the private sector to accept his compromise plan was his use of voluminous data to support his position. Unlike organized labor, the private sector had both money and trained personnel at its disposal. It used these resources to prepare a documented defense of the need and justice of a prior deduction from profits of a percentage of the capital investment of a firm. Margáin decided that the best way to get the private sector to change its position and to accept that of the government would be to prove to its representatives that their arguments and assumptions were erroneous and that the government's position would not prove as detrimental to their interests as they believed.

Margáin then began to meet with the representatives of the private sector, as a group and individually. He made available to them the results of the computer study done by his technical staff. He showed them the amount of profits made by the various kinds of enterprises and applied the private sector's profit-sharing system to his sample of 14,000 firms to prove to them that in many instances there would be few or no profits to distribute to the workers under such a system.

Margáin eventually succeeded in convincing the private-sector representatives that under the system they advocated many workers would be deprived of their right to profit sharing. It was somewhat more difficult, however, to persuade the private-sector representatives that the system proposed by Margáin would not prove extremely detrimental to the private sector. For a time, the private sector representatives refused even to study Margáin's proposal, despite his many meetings with them. Finally, in early December 1963, the private-sector representatives agreed to look at the government's proposal and to give it careful study. Margáin and the technical staff showed them exactly how the government's system would affect business by applying the system to the sample of 14,000 firms.[44]

Once the private-sector representatives had studied the system and its anticipated effects, two of them began to decrease their opposition to the system in general. Their focus shifted to the size of the first two deductions. Significantly, these two representatives had worked in the Income Tax Bureau of the Treasury before becoming active in the private sector.[45]

The government originally favored allowing the two deductions to equal between 10 and 20 percent of a firm's profits. Several of the private-sector representatives, however, claimed that this percentage range was inadequate and demanded a total deduction of 40 percent. After some discussion, Margáin finally agreed to consider allowing the two deductions to total 30 percent rather than 20 percent.[46] Although this figure was still 10 percent less than that advocated by a segment of the private sector, several of its representatives to the commission ceased their opposition to Margáin's system after the government made this concession. The other representatives followed suit a short time afterward.

The private-sector representatives capitulated during the second week in December, several weeks after the labor representatives had accepted the profit-sharing system proposed by Margáin. The labor representatives were somewhat dismayed to learn during informal meetings with Margáin that the government had decided to increase the total of the first two private-sector deductions to 30 percent. The labor representatives did not, however, offer much resistance to Margáin's efforts to persuade them to accept the larger deductions.

At a marathon meeting of the National Profit-Sharing Commission on December 11, 1963, the question of the size of the first two deductions was formally discussed for the first time. The representatives of organized labor stated that they favored two deductions of 10 percent each or a total of 20 percent, although they realized that they would probably have to agree to a 30 percent deduction. The private-sector representatives said they favored a total deduction of 40 percent, but they, too, recognized that they would probably have to accept 30 percent. Margáin suggested 30 percent as a compromise figure. Both parties voted to accept his proposal.[47]

At the same meeting the commission discussed the workers' share of the profits remaining after the three deductions. The labor representatives asked that the workers be given 25 percent of the remaining profits, while the representatives of the private sector favored distributing 15 percent. Margáin suggested a compromise figure of 20 percent. Once again, his compromise was accepted by both sides.[48] This formal discussion of the percentage of the profits to which the workers would be entitled had, as had the formal discussion of the first two deductions, been preceded by separate informal talks between Margáin and the two groups of representatives.

These talks reduced the possibility of increased polarization of the labor and private-sector leaders on the profit-sharing issue and eventually made agreement possible between the two groups of representatives on the National Profit-Sharing Commission. During these discussions, Margáin was able to discover the demands and opinions of each of the groups of representatives, reconcile some potentially conflicting demands and opinions, and persuade each side to give favorable consideration to the position of the technical staff. Therefore, although the meetings of the commission held between March and December 1963 represented the first actual confrontations between the two opposing groups, they were in a sense controlled because Margáin knew from the informal discussions at what points and on what grounds the two sides could be brought into agreement. Only when Margáin was certain that agreement was possible was the specific issue allowed to be formally raised and a vote taken during a commission meeting.

After the size of the deductions and the percentage of profits to be distributed were accepted, the draft of the resolution outlining the entire profit-sharing system was read and voted upon by the representatives of the private sector, organized labor, and the government. The vote was unanimous in favor of accepting the draft. On December 12, the commission held its last meeting, and its members received the personal congratulations of President López Mateos.[49] The resolution of the National Profit-Sharing Commission was published in its entirety in all major Mexican newspapers on December 13.

The commission's unanimous approval of the final resolution was considered an exceptional event by many observers. Previous tripartite commissions, such as those that established the social security system and determined minimum-wage levels, had reached majority decisions, with the government and organized labor usually voting for specific proposals; the private sector, against them. There is no doubt that the government ultimately would have voted with labor against the private sector had the latter refused to accept the government's position, because in late October or early November President López Mateos had authorized Margáin to settle for a majority vote if the private sector persisted in its demand for a prior deduction based on capital investment.[50] The government was willing to make some concessions to the private sector in order to receive its support, but it was not willing to concede a major principle—that the workers in each firm were entitled to receive a share of the firm's annual total profits.

The government, however, definitely preferred a unanimous decision, because that sort of decision would reinforce both the legitimacy of the profit-sharing decision and the unity of the authoritarian coalition. It would also increase the profit-sharing system's chances of success. Members of the private sector were bound to acquiesce more graciously to a profit-sharing system that had been found to be satisfactory by their own representatives. Furthermore, a divided vote (with the private-sector representatives in the minority) might cause loss of confidence within the private sector, which might have had a detrimental (although probably only in the short run) effect on the Mexican economy.

The representatives of the private sector were never averse to voting in the affirmative, provided the costs of so doing were not excessive, since by voting in favor of the profit-sharing system they could reinforce their image as supporters of the Mexican Revolution. Obviously, there was very little possibility that the representatives of the labor movement would withhold support from the profit-sharing system eventually proposed by the head of the National Profit-Sharing Commission.

The commission's unanimous vote can probably be attributed to two main factors. One was the skill of Hugo B. Margáin.[51] The other was the level of development Mexico had achieved by the early 1960s. The private-sector representatives accepted Margáin's proposed profit-sharing system only when he was able to show them exactly how the system would function, only when he could demonstrate that the system would not prove as damaging to the private sector as the representatives feared. He could not have offered sufficiently convincing evidence without a functioning tax system; fairly credible statistical data on taxes, profits, invest- ment, enonomic growth in general, and the size and composition of the labor force; and a pool of experts trained in economics, accounting, and the use of computers. These were just a few of the necessary elements that would have been unavailable in a less-developed country. The private-sector representatives ultimately dropped their original demands and accepted the government's views, therefore, as a result of the availability of credible data that they could also use should they be forced to defend their decision to other members of the private sector.

A less generous interpretation would attribute the unanimous vote to numerous and substantial concessions made by the government to the representatives of the private sector. Although the government refused to concede its major point that a prior deduction based on capital investment was unacceptable, not only the increase in the percentage allowed for the first two deductions but also the deductions themselves represented a concession to the power of the private sector. The third deduction, based on the ratio of capital to labor, would be considered a "sell out" to the biggest, most important firms in Mexico, the obvious beneficiaries of the deduction.

I believe, however, that the deductions allowed by the government did not represent a substantial concession to the power of the private sector. Furthermore, increasing the deduction from 20 to 30 percent can be considered only a minor concession. In the early days of the National Profit-Sharing Commission, Margáin and his technical staff had analyzed the profits of the private sector and had decided that 5 to 7 percent of its total profits could be distributed to the workers without disturbing the investment process. The private-sector representatives had accepted this judgment. The profit-sharing system, which included the two deductions totalling 30 percent of profits, the deduction based on the ratio of capital to labor, and the distribution of 20 percent of the remaining profits to the workers, was calculated to result in the distribution of from 5 to 7 percent of the total profits of the private sector.

The members of the technical staff did not decide to distribute only 5 to 7 percent of the total profits because of pressure from the private sector. The technocrats themselves were concerned about the possible detrimental effects that a greater distribution of profits to the workers might have on the investment process and on economic growth in general. The técnicos were committed to a particular strategy of economic growth that included a dependence on private, domestic capital investment, and this approach was not significantly different from that of large segments of the private sector. Although the percentage agreed upon by the técnicos and the private-sector representatives does represent a compromise, the figure is not too far removed from the percentage first suggested by the técnicos.[52]

The fact that the technical staff to which President López Mateos had delegated decision-making authority and the private sector's representatives shared similar orientations and goals does not mean

115

that the president and his political advisers also shared them. Many of them did not. One of the members of the three-person commission that drafted the 1961 constitutional amendment and the 1962 Federal Labor Law reforms subsequently expressed his dissatisfaction and disagreement with the resolution of the National Profit-Sharing Commission for allowing "a series of deductions that are not permitted either by the [amendment] or the [Federal Labor Law] and for creating a scale that also lacks justification in order to introduce deductions based on the greater or lesser importance that capital and labor represent in each firm." These deductions, he claimed, "have permitted businessmen to reduce the share of profits to a minimum and, on occasion, to destroy the [workers' right to share in profits]."[53]

The orientations and goals of the president and his political advisers were of crucial importance in the deliberation stage and during the early stages of the profit-sharing decision, when the amendment and the Federal Labor Law reforms were drafted. In the later stages of the profit-sharing decision, however, when the technical specifics of the profit-sharing system were elaborated, the orientations and goals of Margáin and his technical staff took precedence. Thus it is not surprising that a participant in the early stages of the decision could disapprove of the details of the profit-sharing system elaborated by the government's technical staff. The ascendancy of the técnicos in the final stages also explains the relative ease with which the private sector's representatives were able to reach an agreement with the government's technical staff concerning the percentage of profits that could safely be distributed. Perhaps President López Mateos originally intended to have the workers receive a larger share of industry's profits. However, his delegation of decision-making power to the técnicos introduced a new set of criteria, and the resulting decision may have differed somewhat from the original conception that he and his political advisers shared.

The Activities of the Private Sector

Both of the two principal groups of interests, the private sector and organized labor, played a role in the later stages of the profit-sharing decision, although it continued to be a reactive role rather than an initiating one. The efforts of organized labor, however, were unimpressive in comparison with those of the private-sector organizations.

The leaders of the private sector devoted their energies throughout 1963 to the collection of information and statistics to support their contention that a deduction from profits based on a firm's total capital investment was necessary to avoid endangering Mexico's economic growth. The leaders of CONCAMIN, CONCANACO, and COPARMEX decided that COPARMEX should take the major responsibility for gathering the necessary data. The findings and conclusions of the COPARMEX study were to be presented to the National Profit-Sharing Commission as unofficially representing the point of view of the private sector in general.

In March 1963, COPARMEX formed a special department to undertake the study, the Department for the Study of Profit-Sharing and Minimum Wages. The department began hiring personnel in May 1963 and by October had acquired a staff of sixty-six individuals. Its finance committee began to solicit voluntary contributions from industrial and commercial firms throughout Mexico and succeeded in collecting 1,500,000 pesos in the Federal District and 500,000 each in Guadalajara and Monterrey.[54] In the meantime, the department was divided into a technical section and an administrative section. The technical section concerned itself with sponsoring economic analyses, gathering statistics, and studying legal matters; the administrative section devoted its energies to building a collection of materials on profit sharing, translating printed matter dealing with the issue, and fostering good public relations.

The department tapped several sources of information, including the libraries of banks and such organizations as the United Nations. COPARMEX also sent several individuals to Europe and South America to gather information on obligatory profit-sharing systems there. COPARMEX also repeatedly pressured its member centers for studies and data concerning specific types of enterprises.

The most important source of information, however, was obtained from questionnaires that COPARMEX, CONCAMIN, and CONCANACO sent to their member chambers and centers throughout the country. There were two types of questionnaires. One was to be sent by the chambers of industry and commerce and employers' centers to their individual member firms. Another was to be answered by the chambers of industry or commerce or the employers' centers on the basis of the questionnaires completed by their member firms. The questionnaires sent to the individual firms asked for information regarding total sales, total number of employees and the total salaries paid to them, depreciation and mortgages, cost of raw materials, income-tax payments, profits and losses, reinvestment, and the total capital investment of the enterprise. The questionnaire that the various chambers or centers were asked to complete included questions regarding the rate of profits obtained by the member firms during the past five years, the percentage of profits reinvested by these firms, and how the term "the capital of a firm" should be defined for the purpose of guaranteeing a fair return on its investment. Other questions pertained to the amount of invested money required in the next five years to assure the adequate development and/or expansion of the particular branch of industry or commerce under consideration, the portion of future needed investment that could be derived from the reinvestment of profits, and the nature of the percentage of profits to be distributed to the workers (i.e., should it vary according to branches of industry or regions or should there be one percentage for all). Information regarding the experience of firms that had previously adopted a voluntary profit-sharing system was also solicited.[55]

Approximately 14,000 questionnaires, accompanied by assurances that the answers would be kept strictly confidential, were mailed to the chambers of industry and commerce and employers' centers in mid-1963. By September, only 2,307 completed questionnaires had been returned. Of these, 955 were from the Federal District and the State of Mexico, 835 were from the other states, and 517 were anonymous. These questionnaires were then processed by IBM of Mexico.[56]

The results of the IBM analysis were included in volume eight of an eleven-volume, 1,733-page study that was published by COPARMEX and submitted to the National Profit-Sharing Commission on behalf of the private sector in October 1963. The comprehensive study followed the ten-point topical outline prepared by Margáin and his technical staff for the commission in February 1963. The study included an analysis of the Mexican economy from 1940 to 1960, suggestions regarding ways of fomenting economic growth, analyses of the obligatory profit-sharing systems of South America and Europe, and a review of the historical antecedents of profit sharing generally and of profit sharing in Mexico specifically. It also provided opinions regarding the fair return to which capital was entitled, the amount of profits that should be reinvested, the definition of "taxable profits" for the purpose of profit sharing, and the percentage of profits that should be distributed to the workers.

The conclusions of the COPARMEX study provided few surprises because they echoed many of the opinions and demands that the leaders of the main organizations of the private sector (i.e., CONCAMIN, CONCANACO, COPARMEX) and the private sector's representatives to the National Profit-Sharing Commission had previously expressed. It was first stated that the time was not yet right to introduce an obligatory profit-sharing system in Mexico and that, as a result, profit sharing might have an adverse effect on the Mexican economy if it were introduced. It was also argued that obligatory profit-sharing systems had failed in all the Latin American countries in which they had been established[57] and that, as a result, extreme care should be taken in the elaboration of the Mexican system. The study reiterated, however, the private sector's acceptance of the principle of profit sharing and stated that the system eventually adopted should allow business firms to deduct 10 percent of their total capital investment from total profits as a fair return on their investment. It further claimed that no one percentage of profits should be guaranteed to workers, but that the percentage should vary between 10 and 30 percent of the profits that remained, if any remained, after taking the 10 percent deduction.[58]

In addition to the eleven-volume COPARMEX study, the efforts that COPARMEX, CONCAMIN, and CONCANACO had expended to encourage their members to present individual studies to the National Profit-Sharing Commission brought some results. Approximately twenty such studies were submitted. They were often fairly sophisticated and included a great deal of economic data and statistics.[59]

The eleven-volume study represented the private sector's greatest effort to influence the commission. The private sector employed another strategy to protect its interests—attempting to persuade the workers to accept a bonus equal to a month's salary in lieu of profit sharing. Although some labor leaders were receptive to this suggestion, Margáin, as has been noted, succeeded in convincing the labor representatives that such a system ultimately would prove detrimental to the workers' interests.

119

When it became clear early in December 1963 that the government was not going to meet the private sector's most important demand (a deduction from profits of a percentage of the total capital investment), the leaders of CONCAMIN, CONCANACO, and COPARMEX resorted to a scare campaign. On December 7, 1963, a full-page advertisement, directed to the attention of "public opinion" and signed by the three organizations, appeared in all the major newspapers in Mexico City. The text of the advertisement stressed the need to guarantee a fair return on capital investment and warned that failure to make such a guarantee could cause capital to leave the country, industries to close, modernization and expansion of businesses to cease, and workers to lose their jobs. Although the advertisement stated that it was intended to inform the public, its real purpose was to mobilize rank-and-file businessmen in support of their leaders and to raise the possibility of massive lack of cooperation on the part of the private sector if the government did not alter its position and allow the kind of deduction demanded by the private sector's representatives. The only effect that the scare campaign had, however, was to cause stocks to decline temporarily a few points. The government remained firm. Shortly afterward, the private sector's representatives to the National Profit-Sharing Commission began to decrease their opposition to the system proposed by Margáin.

On December 9, 1963, CONCAMIN held an emergency meeting, which was attended by some members of CONCANACO and COPARMEX. The statement that was released following the meeting affirmed the private sector's demand for a prior deduction based on capital investment.[60] The newspapers, however, did not print the fact that the private sector's representatives to the National Profit-Sharing Commission were no longer unanimously opposed to the profit-sharing system proposed by Margáin. The December 9 meeting was used, therefore, to explain Margáin's proposed system, to indicate that it would not prove detrimental to the private sector, and to mobilize support for it in the form of a vote of confidence. Once the confidence was expressed, the private sector's representatives to the commission agreed to restate their original demand. In the event that Margáin remained intransigent, however, they agreed to accept his system instead. If the private-sector representatives to the commission had not received a vote of confidence at the meeting, they would have been obligated to vote against Margáin's proposed profit-sharing resolution.[61]

It might be noted that the high degree of unity that the private sector manifested throughout the profit-sharing decision is not always typical of its behavior. On certain issues the major organizations, CONCAMIN, CONCANACO, and COPARMEX, take opposing stands although their respective members remain united. On still other issues not only are all the major organizations divided, but each organization also experiences extreme internal disunity and competition. The latter case has been described by Rafael Izquierdo, with reference to the issue of protectionism: "At the institutional level the private sector's bargaining position is limited by the diversity of its interests and its consequent inability to present a united front in the agencies charged with the administration of policy. Each entrepreneur is concerned with imposing his own point of view and resolving his own problems . . . Private enterprise as a group has no common interests with regard to protectionist policy."[62]

The unity of the private sector in the case of profit sharing can be explained by both the nature of the issue and its predicted repercussions. Profit sharing was perceived as a "redistributive" issue by the members of the private sector, who viewed it as a mechanism for diverting income from the private sector to organized labor.[63] It was also a multisector issue because the members of *all* the private sector organizations, whether industrial or commercial, were to be obligated to share their profits with their employees.[64] Decisions that are both redistributive and multisector (e.g., profit sharing) produce *active* cohesion within and among the major organizations (e.g., CONCAMIN, COPARMEX, CON-CANACO) that represent the class that perceives its interests *threatened* (e.g., the Mexican private sector).[65]

The Activities of the Labor Groups

The activities of the leaders of organized labor during the 1963 phase of the profit-sharing decision were extremely limited. They did not undertake a study comparable to that of the leaders of the private sector. The private sector submitted nearly twenty studies, excluding the eleven-volume COPARMEX study, to the National Profit-Sharing Commission; the labor groups presented only six papers.

The labor leaders' only other activity involved reacting against what they perceived as efforts of the private-sector leaders to deprive organized labor of its right to receive a just share of industry's profits. When the private sector's representatives to the National Profit-Sharing Commission attempted to substitute a bonus equal to one month's salary for profit sharing, the labor leaders, in part as a result of conversations with Margáin, denounced the private sector's maneuver to the newspapers. The private sector's December 1963 advertisement provoked the labor leaders into sponsoring an equivalent advertisement that contradicted all the assertions of the private sector and presented the viewpoint of the labor movement.[66] Once the private-sector representatives dropped their main demand and accepted the profit-sharing system proposed by Margáin, however, the labor representatives ceased their attacks. The "spirit of mutual comprehension, conciliation, and tranquility," which President López Mateos in September 1963 had claimed characterized labor-employer relations on the National Profit-Sharing Commission, became a reality.[67]

Only during the 1963 phase of the profit-sharing decision were the labor leaders united. They united only because the government had forced them tó confront their opponents on the National Profit-Sharing Commission. Apparently, redistributive, multisector policies will produce neither active nor passive cohesion among the major organizations of the class that benefits from the decision unless their leaders are forced to confront the leaders of the organizations representing the class whose interests are threatened by the decision.

Enforcement of the Profit-Sharing Decision

The publication of the resolution outlining the profit-sharing system that had been unanimously agreed upon by the members of the National Profit-Sharing Commission resulted in expressions of satisfaction and pride on the part of all concerned groups. CONCAMIN, CONCANACO, and COPARMEX praised the "equilibrium" of the resolution, its "high technical content," its recognition of the need to sustain private investment and of the private sector's right to a return on its investment.[68] The Bankers' Association broke its guarded silence in order to offer its congratulations to all parties.[69] The head of the CTM stated that "the working class is exuberantly celebrating because of the profit-sharing resolution, which we can consider to be our greatest conquest of the last twenty-five years . . ."[70] Even the FOR, a more radical labor organization, expressed its "full and complete support" for the resolution and called it "a revolutionary measure, since it will not affect the class struggle but . . . will give impetus to industrialization."[71] Other voices in the approving chorus included those of the head of the PRI and the editors of Mexico City's major newspapers.

It is not very surprising that the spokesmen for the private sector approved of the resolution. When President López Mateos had first announced his decision to establish an obligatory profit-sharing system, the members of the private sector were fearful that such a system would prove extremely detrimental to them. The final resolution of the National Profit-Sharing Commission showed their fears had been unjustified. The resolution allowed two 15 percent deductions to be taken immediately from taxable profits. This 30 percent deduction represented "interest on invested capital" and a "reinvestment" incentive. From the remaining 70 percent of taxable profits, a deduction based on the ratio of capital to labor (total wages) was allowed. This deduction was to vary from between 10 and 80 percent of the total taxable profits remaining. The most highly capitalized industries (i.e., those with the largest capital/labor ratio) were allowed the largest deduction. After all these deductions had been taken, 20 percent of the remaining profits were to be distributed among the workers. As a result of this formula, the private sector would at no time have to share more than 12.6 percent of its total taxable profits. The minimum percentage it would have to share was 2.8 percent. Furthermore, the spokesmen for the private sector were also representatives of Mexico's largest and most highly capitalized firms. They had originally expressed great concern over the impending profit-sharing system, because they had feared that it would interfere with their investment plans. The final profit-sharing resolution, because of the deduction based on the ratio of capital to labor force, dissolved their fears, since it provided for the most highly capitalized industries to share the minimum of 2.8 percent of their profits.

The owners of Mexico's smaller businesses (i.e., those with annual incomes between 120,000 and 300,000 pesos) did not applaud the profit-sharing resolution. The resolution stipulated that, as a result of the difficulty of accurately ascertaining taxable profits for these small firms, for the purposes of profit-sharing taxable profits were to be considered the equivalent of 17 percent of a firm's gross annual income. In reality, however, many small businesses do not earn profits equal to 17 percent of their gross annual income. In addition, because these smaller firms are more labor-intensive than the larger enterprises, the deduction to which they are entitled based on the capital/labor ratio is substantially smaller than the deduction allowed the more highly capitalized firms. The owners of small businesses, therefore, were the hardest hit by the resolution. They had not been represented on the National Profit-Sharing Commission; and they were not influential in CONCAMIN, CONCANACO, or COPARMEX.

The fact that the spokesmen for organized labor applauded the resolution is also not surprising. Although the workers would not receive as great a share of the profits as they had anticipated, they would still receive a percentage of their employers' profits, whereas, prior to the resolution, all profits had gone to the employer. Furthermore, the highest paid workers, a category that included the spokesmen of the labor movement, would receive the most money, since under the system adopted half the profits were to be distributed in proportion to a worker's salary. Misunderstanding also contributed to labor's enthusiasm. Some labor leaders were under the impression that the workers were to receive 20 percent of the profits and did not realize that as a result of the various deductions, the actual percentage that they would receive would range between 2.8 and 12.6 percent.[72] In terms of the workers' salaries, it was estimated that their additional income would range between a low of one week's to ten days' pay and a high of one month's pay.[73]

It was not long, however, before organized labor's expressed satisfaction with the resolution noticeably decreased. The cause was not the lower-than-expected share of profits to which labor was entitled, but the attempts by employers to evade the profit-sharing law. In June 1964, Fidel Velázquez, the head of the CTM, stated that a minority of the industrialists were complying with the law, but that "the immense majority [were] committing frauds and violations." He added that of 40,000 businesses in the country, fewer than 3,000 had shared profits.[74] A short time later, the CTM leaders asked all member organizations for information regarding the amount of money their workers had received as profit sharing and the names of firms that had not distributed profits. By doing so, the leaders hoped to obtain hard data with which to support their accusations.[75] Charges of evasion on the part of employers were made at CTM assemblies throughout 1964, and early in 1965 the CTM began to consider using the threat of a strike to force compliance with the profit-sharing law.[76]

The CTM's charges were definitely valid. Profit sharing was based on the amount of taxable profits declared in tax returns, and the Mexican government lacked the resources and capabilities necessary to verify each firm's computations. The government had more control over the larger enterprises in this respect because they were less numerous and could therefore be kept under closer surveillance. The middle-sized firms, therefore, risked less by inaccurately reporting their taxable profits, and they took advantage of the situation. The larger firms, however, could decrease their relative disadvantage by hiring the best accountants to aid them in their search for legal means of reducing the amount of their taxable profits. One such method involved revamping the organizatonal structure of a company by converting some of its directors, who were not eligible to share in profits, into *empleados de confianza* (high-level assistants), who *were* eligible. Since half the share of profits to be distributed was directly proportional to an employee's wages, by purposely inflating the salaries of the empleados de confianza the amount of total profits to be shared among regular employees was reduced. Other methods of decreasing the amount of profits to be distributed included merging a company that was highly profitable with one that consistently lost money and inflating business expenses by undertaking excessively expensive advertising campaigns, ambitious plant and office renovation projects, and the like.

If it was difficult for the government to verify the accuracy of employers' declarations of taxable profits, it was impossible for the workers to do so. The Federal Labor Law obligated employers to "make known to their workers" the declaration of taxable profits submitted to the Income Tax Department of the Treasury. The law did not, however, require employers to provide the labor unions with copies of the various supplements or *anexos* that showed the deductions that had been made from gross profits in order to arrive at the taxable-profits figure. It was therefore impossible for the accountants of the labor unions even to estimate the accuracy of the taxable-profits figure (and, by extension, the amount of money due the workers), for the necessary information was not available to the workers.[77]

By 1965, concern over circumvention of the profit-sharing law had increased to a sufficiently high level that CTM leaders began demanding an end to the evasion. In September 1965, their representatives in the Chamber of Deputies submitted a project for reforms to Article 123 of the Constitution. The reform project included a provision that workers be given access to the anexos that showed the gross profits of a firm and the amounts deducted from the gross profits in order to arrive at taxable profits. Another clause stated that partial or total noncompliance with the profit-sharing law by a firm should be considered a sufficient legal cause for a strike.[78] The CTM also began demanding the creation of a special tribunal to deal with violations of the law and the resulting conflicts.[79] By 1967, the CTM demands had been adopted by the Congreso del Trabajo, the new labor group that resulted from a merger of the BUO and the CNT in 1966 and included the entire organized labor movement. The leaders of the new group presented the demands to the secretary of labor in November 1967. At approximately the same time, the labor leaders began to focus their efforts on influencing the final form of the totally new Federal Labor Law the Díaz Ordaz government was in the process of drafting, in the hope that the final draft would include their demands.

The government, in the meantime, was aware that the profit-sharing system was producing new kinds of problems. In April 1964, therefore, it created a Department of Profit Sharing and Minimum Wages within the Ministry of Labor to deal with questions, disputes, and violations regarding the profit-sharing and minimum-wage laws. The department was given the power to present legal opinions or judgments, but it was given powers of enforcement only in cases involving firms under federal jurisdiction.[80] When an enterprise that was not under federal jurisdiction was involved, the matter under consideration was forwarded to the appropriate state. There it was dealt with by a state junta of conciliation. During the first years, the department annually received about 250 complaints, of which less than one-third came under its jurisdiction. It also averaged 200 consultations per year.[81] By 1971, the number of complaints received had decreased to 16, and the consultations had declined to 107.[82]

In the spring of 1964 the government also established a Profit-Sharing Department within the Income Tax Department of the Treasury. This new department's functions overlapped somewhat with those of the one formed in the Ministry of Labor. Both answered questions and gave legal opinions regarding disputes arising from the profit-sharing legislation. The power of the Profit-Sharing Department of the Treasury was not, however, limited to enterprises under federal jurisdiction. It also had access to all income-tax declarations filed with the Mexican Treasury and could use the Treasury's auditing staff to verify the accuracy of an employer's profit-sharing calculations when workers challenged them. The department at first handled between 400 and 500 cases each year. Thereafter, its case load increased approximately 10 to 15 percent annually. In the first years of its operations, the majority of the cases were resolved in favor of the employer, but when the department began to improve its fiscal techniques and expand and upgrade its staff the number resolved in the employer's favor began to decline.[83]

In July 1967, the government announced the creation of another profit-sharing department within the Treasury. Unlike the Treasury's other profit-sharing department, which heard and resolved disputes and checked violations, the new department was to educate the workers regarding their rights under the profit-sharing law, teach them how to calculate the amount of profits due them, and provide free counsel when needed.[84]

The creation of these three profit-sharing departments helped resolve some of the problems that arose after the introduction of profit sharing and contributed in a small way to the reduction of evasion on the part of employers. However, the problem of the workers' lack of access to the anexos that showed how the employer had figured taxable profits remained unresolved.

Early in 1968, when the government began to circulate the draft of the totally revised Federal Labor Law to obtain opinions, it became evident that the government was planning to act. The draft included the provision that "the employer . . . shall make known to his workers his annual income-tax declaration, accompanied by the 'anexos' and documents that the laws require."[85] The "ante-proyecto" of the new labor law included two other innovations. In the future, the National Profit-Sharing Commission was to

127

determine the percentage of profits to be distributed among the workers "without making any deductions or distinguishing among types of industries."[86] Furthermore, lack of compliance with the profit-sharing laws was added to the list of legitimate causes for declaring a strike.[87]

Despite strong opposition from the private sector, all these modifications were included in the version of the Federal Labor Law President Díaz Ordaz sent to Congress on December 13, 1968. When he learned that the changes had been retained, the president of CONCAMIN, Prudencio López, charged that the government had ignored "almost all of the objections presented by the private sector" and asserted that the new law "destroys the profit-sharing formula, establishing instead a system that will result in unnecessary unrest." He also claimed that the section regarding strikes was "unwarranted and dangerous tyranny against employers."[88] Unlike the leaders of the private sector, the labor leaders expressed satisfaction with the proposed reforms that related to profit sharing.[89]

After additional studies by a special congressional committee, the proposed labor law was finally presented to the full Chamber, which approved it on November 12, 1969. The Chamber's principal modification was to place a ceiling on the salaries of empleados de confianza for the purpose of limiting their share of profits, thereby eliminating loopholes in the original profit-sharing law.[90] The revised law was approved by the Senate on December 2, 1969 and took effect on May 1, 1970.

It is significant that several of the innovations contained in the new Federal Labor Law, specifically those relating to the right to strike and the provision of anexos to the workers, originally had been proposed by the CTM as part of the reform project it submitted to the Chamber of Deputies in September 1965. As in the case of the original profit-sharing reforms, the government once again had chosen to ignore a group's demands until a time it deemed appropriate to act. The government had no doubt realized that evasion would occur as soon as the profit-sharing system began to function. It probably believed, however, that to attempt to eliminate loopholes immediately after introducing a new system would prove politically unwise. It therefore allowed some time to pass before taking steps to perfect the functioning of the profit-sharing system. Its earliest efforts were calculated to disturb the employers as little as possible (e.g., the creation of departments to receive complaints and provide answers to questions). By 1968, after the profit-sharing system had been functioning nearly five years, the government finally began to correct several of the more serious problems, especially the workers' inability to verify the employers' statements regarding taxable profits.

Future government actions will undoubtedly be directed toward further increasing the government's ability to protect the workers' right to receive a share of a firm's profits. By so doing, the government simultaneously increases its ability to prevent tax evasion by employers. It is also likely that when the second National Profit-Sharing Commission is constituted some time after December 1973,[91] it will take steps to augment the amount of profits distributed among the workers. The government has never done a systematic study of who receives money as a result of the profit-sharing system or how much money each worker receives (if it has done such a study, it is a well-kept secret).[92] A recent pilot study of 83 firms by a COPARMEX economist, however, indicates that although the amount of money an average worker received between 1966 and 1970 as his share of profits increased from 1,191 to 1,717 pesos (or from U.S. $95.28 to U.S. $137.36), the increase did not keep pace with the corresponding increase in private-sector production and sales.[93] The labor leaders have been making exactly this argument for several years, based on rough comparisons between the rate of growth of production in the private sector and estimates of the total number of pesos distributed each year.[94] These data, the new labor law's elimination of deductions previously allowed the private sector,[95] and worker discontent resulting from recent high rates of inflation—all suggest that the government will move toward increasing organized labor's income from the profit-sharing system.

The Dynamics of the   6
Authoritarian
Decision-Making Process

The profit-sharing decision provides insight into two important aspects of the Mexican decision-making process: the procedural aspect or the behavior patterns that characterize it and the identification of the individuals or groups who benefit in a substantive way from the decisions made.

Although my conclusions are based mainly on the evidence gathered from the profit-sharing study, I have also drawn upon several less detailed works on the Mexican decision process. These studies include Schmitter's and Haas's monograph on Mexico's decision to join the Latin American Free Trade Association (LAFTA), McCoy's essay on Mexican population policy making, Benveniste's examination of the educational planning process, and Greenberg's analysis of the functioning of the Ministry of Hydraulic Resources.[1] The congruence between the findings in these works and those derived from the study of the profit-sharing issue lend support to the contention that it is possible to speak of the existence of a uniquely Mexican decision-making process.

Because throughout this work Mexico has been treated as a type of authoritarian regime, however, one should also be able to find decision procedures similar to those identified in Mexico in other authoritarian regimes. Although case studies of the decision process in authoritarian regimes are rare, Anderson's *Political Economy of Modern Spain* offers conclusions with which the Mexican findings can be compared. In this book Anderson examines the policy-making process of Spain, generally considered a prototype of an authoritarian regime, focusing on the stabilization program and liberalization policy of 1959 and the experiment with national economic planning of 1963. His conclusions parallel those based on the Mexican profit-sharing decision and reinforce the idea that there is a decision-making process common to authoritarian regimes.

Patterns of Decision Making

In Mexico, the decision-making process is formally initiated by the executive. In the first stage, the president commits himself to a particular idea that he may or may not have originated. The actual origin of the idea is not important, however. What matters is the president's commitment to it.[2]

All studies of the Mexican decision process emphasize the importance of executive initiative. The profit-sharing decision originated "at the top" and came as a surprise to the affected groups. President Echeverría's reversal of the government's position on family planning followed a similar pattern. The LAFTA decision was made by high-level government technocrats. Educational planning, according to Benveniste, took place "silently and secretly behind the doors of the Ministry of Education and the Bank of Mexico,"[3] and the process was controlled by the executive. And the study of the Ministry of Hydraulic Resources concluded that "over-all planning, including the important aspect of site selection, is an executive prerogative."[4] As described by Anderson, the Spanish situation is comparable, for both the stabilization and economic planning programs originated within the executive.

The president's commitment to a particular course of action rarely is the result of direct pressure by concerned groups. The co-optation of group leaders (which reduces the autonomy of interest groups) and the low level of mobilization of the rank-and-file membership make it difficult for groups to pressure the

131

executive. Patrimonial rulership, with its divide and rule tactics, further isolates the president from group pressures by decreasing the possibility of unification among the groups for purposes of increasing their leverage on the executive.

The fact that the profit-sharing decision did not result from direct group pressures therefore is not unusual. Mexico's decision to join LAFTA could not be attributed to group demands. In fact, after the decision was announced, the government's technocrats had to devote considerable time to convincing the private sector that the decision was in its interests. Likewise, President Echeverría's abrupt termination of his government's opposition to family planning could not be correlated with strong and effective pressures by concerned groups. In the educational planning process, the views of relevant and often interested groups like SNTE (the Teachers' Union) were often ignored because the SNTE lacked the ability to force planners to take its views into account. Of all the cases mentioned, however, interest-group pressures were perhaps least relevant to the decision-making process within the Ministry of Hydraulic Resources, because the greater technical expertise required insulated the decision process even more from public pressures and demands. The Spanish decision-making process also approximated the Mexican pattern. Specifically, Anderson notes the relative autonomy of the policy-makers, who did not seem to have "to bend their designs to the will of interested publics . . ."[5]

As a result of their limited autonomy, interest groups in an authoritarian regime play an essentially reactive role in the decision process, either supporting or opposing some or all aspects of a decision. If they oppose a decision, direct criticism of the president is avoided, since such criticism would imply insubordination. Criticism is therefore focused upon specific aspects of the decision, upon the procedure followed in making the decision, or upon a subordinate of the president who has been closely identified with the decision.

In the case of profit sharing, for example, the labor groups were quick to offer their support for a decision about which many of them were still ambivalent, while the private sector reacted by criticizing the secrecy surrounding the decision rather than the decision itself. The study of population policy making highlights the reactive role of the Mexican Catholic Church, which opposed artificial contraception until *after* the government altered its position regarding family planning. The Church then "reinterpreted" the papal encyclical *Humanae Vitae* and decided it favored family planning in the "emergency" situation facing Mexico.[6] The Spanish study also emphasized the reactive role of interest groups. Like their Mexican counterparts, the Spanish labor leaders "merely adapted [themselves] to a design prepared by others" and worked to mobilize support among their rank and file for the government's new policy.[7]

In the absence of direct group pressures, the president has relatively great latitude in deciding what course of action to pursue. His freedom of action helps to explain the rather sudden shifts in policy that are evident in some of the case studies. The most striking example is President Echeverría's abrupt reversal of government policy toward family planning, but the sudden public commitment to the establishment of a profit-sharing system or to comprehensive educational planning are also a partial result of such presidential latitude. Anderson's study also stressed the "wide range of alternative economic techniques and measures" available to Spanish policy-makers[8] as well as the flexibility and adaptability of the policy-makers, who were relatively free from group constraints.

The relative freedom from interest-group pressures means that the president's commitment to a decision can usually be attributed to less specific factors such as the goals, interests, and values of the president; the international situation; the sudden availability of a new solution to a persistent problem; the advent of fashionable ideas from abroad; and the president's desire to increase his support and legitimacy by implementing a goal of the elite consensus. In an authoritarian regime, the inability of groups to exert significant pressure means that such factors assume greater importance in accounting for presidential commitment to a particular decision than they would in a regime in which groups are relatively autonomous and highly mobilized.

In the case of profit sharing, a number of factors, including López Mateos's background as labor minister, the pressure of the Cuban Revolution, the government's improved tax collection capabilities, and the desire of the president to implement a portion of the 1917 Constitution account for the decision. The LAFTA decision, which occurred during the same period, also was in part the result of the need of the Mexican leaders "to demonstrate their own brand of progressivism in the face of the more radical Cuban challenge . . ."[9] McCoy's explanation for the change of attitude toward family planning points out such factors as the influence of the discovery and debate of the "population explosion" in other countries and the increased availability of effective contraceptives in Mexico. Finally, the Spanish liberalization and stabilization programs were to a substantial degree influenced by the successful application of these techniques by French technocrats.

Sometimes, however, the president's freedom of choice is restricted by indirect pressures from groups within the polity that have patiently watched as the president consistently neglected their demands and interests. The president may finally decide that such groups can no longer be kept demobilized by promising future rewards, co-opting group leaders, or replacing uncooperative leaders with more cooperative ones. He may therefore make a decision that he feels will placate these groups, ideally without sacrificing any of his other goals.

In the case of profit sharing, the 1961 reforms of the labor article of the 1917 Constitution can be attributed in great part to indirect pressures of this kind. Although none of the other Mexican decisions studied seem to have resulted from the desire to placate a particular neglected group, they can be attributed to diffuse pressures of a slightly different nature. The policy favoring family planning, for example, represented an attempt to forestall a political and economic crisis occasioned by a rapidly increasing population, while the new emphasis on educational planning was to some degree related to the 1968 student uprisings. In Spain, the stabilization, liberalization, and planning programs were linked to a desire to revive the economy, which was "virtually bankrupt and registering chronic balance of payment deficits,"[10] a situation that could have undermined political stability.

On rare occasions, however, the president may overestimate his ability to prevent the mobilization of opposition, usually because he has faulty or insufficient information. The dissatisfied groups may become highly mobilized and determined to improve their situation by using their only recourse—disruptive or violent activity. When such disruption occurs, the regime first will repress the leaders of the demonstrating groups and replace them with individuals who are willing to cooperate and who are able to demobilize their followers. Once order has been restored, the president may commit himself to a decision that will benefit the previously neglected group. The 1957-1959 labor disturbances, for example, resulted from the persistent neglect of organized labor by Mexican presidents. The profit-sharing reform of 1961 was part of a "pay-off" to organized labor once the labor movement had been demobilized.

Once the president has committed himself to a particular course of action, a phase of deliberation follows. A very small number of people participate in the initial deliberations, which are not made public. If a recommended decision is expected to provoke opposition, the president and his advisers will agree to make the decision, provided they believe they will be able to demobilize its opponents during subsequent phases of the decision-making process and, even better, to convert them into supporters. In addition, they must be able to mobilize the sentiments of their supporters (who are relatively quiescent) in order to balance the opposition that the decision is expected to generate.

In the case of profit sharing, the president apparently decided that he would be able both to demobilize the business groups and mobilize the support of the labor groups. The president seems to have made a similar assessment with regard to the decision to join LAFTA, which, like the profit-sharing decision, met with considerable resistance from the private sector, and the family planning decision, which produced dissatisfaction among strong Catholics in general and within the Catholic Church in particular. In Spain, despite the fact that the labor unions were to bear the brunt of the government's stabilization program, the government anticipated and ultimately received the support of the union leaders and their followers.

The mobilization of support for a decision may prove difficult for several reasons. First, the potential beneficiaries of the decision are divided among themselves as a result of patrimonial rulership, which encourages vertical authority relationships and discourages horizontal ties within and among groups. In addition, the intended beneficiaries usually have a low level of mobilization. They seem to have little incentive to unite against their opponents, for they may feel, as did the labor groups in the case of profit sharing, that the president will safeguard their interests. Second, the supposed beneficiaries of a decision may not regard the decision as serving their interests. Several labor leaders, for example, perceived the profit-sharing decision as threatening to labor's class consciousness. In the LAFTA decision, the private sector did not immediately recognize the potential benefits that it might reap as a result of Mexico's membership in the association. The Spanish entrepreneurs were not certain that the government's new economic policies would work to their benefit. Finally, it is possible that a decision the government wishes to make may be viewed as detrimental by all affected groups. If the regime must make an excessive effort to mobilize support for a controversial decision, it is unlikely that the decision will be made. In all the cases under consideration, the executive apparently concluded that sufficient support could be mobilized without paying too high a price.

Once the Mexican president and his advisers are in agreement regarding the wisdom of making the decision, the president publicly associates himself with it by making a formal announcement or an executive-sponsored legislative proposal, or both. All important decisions are formally initiated by the president, and the president both claims and receives full credit for the decision, whether or not the idea for the decision was originally his. Because of the patrimonial nature of staff arrangements, all individuals who participate in the decision-making process supposedly do so at the president's will and serve in the capacity of his subordinates. In return for receiving the delegated power to serve, they attribute all credit for their accomplishments to their patrimonial leader, the president. The authority of the president and, indirectly, the integration of the authoritarian coalition, which is headed by the president and composed of the co-opted group leaders and their followings, are thus reinforced. Another effect of this convention is the extreme difficulty (if not impossibility) of discovering who is actually responsible for initiating the decision. All inquiries regarding the origin of the profit-sharing idea, for example, were answered by attributions of credit to President López Mateos.

If the president's decision should happen to correspond to a previously expressed demand, as the profit-sharing and the family planning decisions did, it cannot be directly attributed to the demand, which plays an informational rather than a causal role in the decision process. The time lapse between the initial expression of the demand and the president's decision is too great for the decision to be interpreted as a response to the demand. The CTM first began to campaign for the implementation of the Constitution's profit-sharing clauses in 1951; the clauses were implemented ten years later. Groups favoring a governmental program to help limit population growth began expressing their views in the early 1960s; President Echeverría's reversal of government policy occurred a decade later. Furthermore, the president's decision in both cases was not announced during the period in which the groups in question were expressing their demand most vociferously. In the case of profit sharing, active campaigning for the idea reached its height in the early 1950s; family planning advocates were most vigorous in the mid-1960s. The timing of the decision emphasizes that the president's choice is dependent upon his own will rather than upon that of his subordinates. In this way, the patrimonial image of the president is reinforced, and the authoritarian coalition remains integrated.

A decision made in response to a violent outbreak follows a similar pattern. A substantial period of time will be allowed to elapse between the outbreak and the decision so that the latter will not be interpreted as a response to the former. The labor unrest of the late 1950s, for example, was not followed by new legislation benefiting the labor movement immediately, but approximately four years later. The delay not only reinforces the president's patrimonial image, but avoids giving the impression that violent disruption is rewarded. To give such an impression would encourage a higher level of mobilization and would indirectly threaten the stability of the authoritarian regime.

If the announced decision is not expected to provoke significant opposition, the problems of demobilization of the opposition and mobilization of support do not arise. The announcement of such a noncontroversial decision is accompanied by or is soon followed by the submission to Congress of a legislative proposal. The proposal is drafted within the relevant government ministry and the relevant interest group leaders are only consulted if the government feels they can provide the ministry with information it lacks or if the interest group leaders request consultation to communicate opinions or suggestions they would like the president to take into account. Obviously, in the case of noncontroversial decisions, no effort to shroud the government's activities in secrecy is made. Congressional approval of the legislative proposal is never in doubt, even if the decision is controversial, since congressmen owe their position to the authoritarian elite rather than to their supposed local constituents.

With a potentially controversial decision the procedure is quite different. In Mexico, economically redistributive decisions like the establishment of a profit-sharing system, as well as politically redistributive decisions, fall into this category. Such decisions are considered capable of mobilizing substantial opposition that might prove destructive to the authoritarian coalition. Thus, a great effort is made to avoid the mobilization of opposition and to demobilize the decisions's opponents.

One way of demobilizing the decision's opponents involves the use of secrecy. Unexpected announcements find potential critics unprepared. For this reason, no consultations with interest-group leaders occur before the decision is announced. This strategy was used in both the profit-sharing decision and the sudden reversal of the government's position on family planning. Furthermore, few specific details regarding the decision are offered, since the less that is revealed, the fewer aspects of the decision are open to challenge. Another reason for avoiding detail is that a decision may represent primarily a political commitment, and the technicalities have not been thoroughly worked out. Both reasons explain why the 1961 amendment to Article 123 of the Constitution contained few details regarding the nature of the profit-sharing system to be established. In the educational planning process, according to Benveniste, announced plans also were vague because they represented essentially political commitments. A final reason for the absence of detail relates to the fact that the Mexican regime is in part based on "constitutionalism." The president is thus obliged to obey the letter, if not the spirit, of the law. The more general a law is, the more the autonomy of the Mexican president is preserved, since he has more latitude to interpret it as he sees fit. Vague laws are thus one way of reconciling legal legitimacy with patrimonial rulership.

The presentation of the unanticipated legislation is also timed so as to allow minimal opportunity for public debate. In Mexico, such legislation is sent to Congress during the final days of its annual session. To debate it at that time could delay action on it until the following session, a result that congressmen, as "subjects" of the patrimonial leader, seek to avoid.

Controversial legislation is usually also preceded and followed by legislation that benefits other sectors of the population. In this way, interests not benefited by the legislation will not feel neglected by the regime. The profit-sharing legislation thus was preceded and followed by legislation that benefited the military, government bureaucrats, and rural groups.

Controversial legislation is also justified in terms of the "revolutionary" ideology, which guarantees that its legitimacy is unassailable. The use of the prevailing ideology in this way has been called "the mobilization of bias."[11] In Mexico, the mobilization of bias against the interests of private groups is a frequent tactic of the government. It also serves to reinforce the integration of the authoritarian coalition by emphasizing the nation rather than its component parts. Because profit sharing was an unimplemented provision of the 1917 Constitution, the decision to initiate it was in principle unassailable and reinforced the legitimacy and integration of the Mexican regime. In the case of the decision favoring family planning, the mobilization of bias limited the ability of the Catholic Church, an "unrevolutionary" institution, to challenge or question the president's decision.

The main device the authoritarian elite uses to demobilize its critics is the incorporation of the malcontents into the decision-making process. Incorporation only occurs, however, *after* the initial vague version of the legislation has been approved by Congress or, if no legislation is involved, after the vague version of the decision has been publicly announced. Incorporation thus follows the political commitment or the "decision in principle." The opposition groups then are invited to participate in the elaboration of a more specific law that will implement the initially unspecific legislation. In the case of profit sharing, the representatives of the private sector and organized labor were incorporated into the decision-making process after the 1961 constitutional amendment had been passed but before the drafting of the more specific Federal Labor Law reforms. Similarly, consultations involving educational planning took place after the elaboration of the basic plan. By delaying the incorporation of group representatives, the president avoids subjecting his decision in principle or his political commitment to discussion. Participation is thus confined to the elaboration of technical details and implies the acceptance by the groups of the president's political commitment. In his study of

Spanish policy making, Anderson describes this procedure as one that makes "political or potentially political issues into technical ones."[12]

Incorporation involves consultations between a representative of the regime and the leaders of relevant and officially recognized groups. The group leaders are not allowed to express their views in each other's presence. The representatives of the private sector and those of organized labor each met separately with the secretary of labor to discuss profit sharing. This procedure does not permit differences of opinion to escalate and keeps the level of mobilization low. The patrimonial divide-and-rule consultations also reinforce the ties of each of the groups with the center. They also avert horizontal clashes among the groups, which would mobilize the groups and weaken both the coalition and its unifying central agent. Such consultations in no way obligate the authoritarian leader to accept all the group leaders' suggestions. In Anderson's words, "representation is consultative and not controlling."[13] For the president to accept all suggestions would imply that the modern patrimonial ruler was subject to the will of his subordinates.

The results of the consultations depend on the demands of the respective groups, the extent of their autonomy, their level of mobilization, and the strength of the president's commitment to the decision. The most highly mobilized and autonomous groups will receive the greatest concessions. In the profit-sharing decision, more concessions were made to the private sector than to the labor movement. In the case of the Spanish stabilization program, the business groups also fared better than their more dependent labor counterparts. In the educational planning process, the combination of the government's commitment to a particular plan and the relative weakness of the Teachers' Union often produced a situation in which the union's views either were not solicited or were solicited and then ignored.

In the case of highly controversial decisions, incorporation often includes a government-mediated confrontation of opposing groups during the *final* phase of the decision process. The potentially destabilizing and disintegrating effects of interest-group confrontation are thus avoided, since the residual problems dealt with during the final phase are highly specific and technical.[14] Groups that are deficient in technical expertise, such as the Mexican labor unions, defer to the regime's technocrats. Groups that possess or can purchase the services of technical experts, such as the private sector, defer to their experts. The remaining details are therefore resolved by technocrats, whose outlook and values often converge, rather than by politicians or interest-group leaders. The focus for interest-group confrontation in Mexico is usually a government-sponsored *ad hoc* commission headed by a trusted presidential subordinate to whom the president has delegated substantial decision-making authority. The *ad hoc* commission is the chosen vehicle because the president does not wish to risk a permanent appropriation of his decision-making authority. The National Profit-Sharing Commission came into existence in early 1963 and was dissolved in December 1963, after it had elaborated a specific profit-sharing system.

The head of the *ad hoc* commission is a technocrat, not a politician. The choice of a technocrat reinforces the idea that the commission is to deal with technical rather than political questions. Hugo B. Margáin, the head of the National Profit-Sharing Commission, had been a high-level bureaucrat and had not been active in party politics.

The government-mediated confrontation on the *ad hoc* commission has political as well as technical functions. Without such a confrontation, the beneficiaries of a decision, who are usually not highly mobilized, have no incentive to unite in support of the authoritarian leader, because they regard his decision as a gift that they did not have to work actively to obtain. The authoritarian leader, however, may find it difficult to avoid making concessions to the more highly mobilized opponents of the decision unless the beneficiaries are united in his support. In the case of the profit-sharing decision, the confrontation served to unite the leaders of the faction-ridden labor movement and enabled the regime to avoid conceding its major point—that no deduction from profits based on a percentage of a firm's total capital investment should be allowed. An additional function of the confrontation is the reconciliation of the leaders of the recipient groups (e.g., the representatives of organized labor) with the leaders of the groups whose interests suffered as a result of the decision (e.g., the representatives of the private sector).

Once the specifics of a decision have been elaborated, attempts to perfect their enforcement will be incremental so as not to mobilize further the objecting groups and destabilize the system.

Minor adjustments or corrections will be made first. The earliest attempts to enforce the profit-sharing system, for example, involved the creation of offices to increase the workers' and employers' understanding of the mechanics of the system. Later reforms increased the workers' ability to verify the accuracy of their employers' declarations concerning the amount of profit. The poor initial enforcement procedures may be regarded as performing a demobilization function, for opposing groups will reduce their opposition if they believe that a decision to which they object will not be stringently enforced.

Who Gets What and Why

The profit-sharing decision produced no clear winners or losers. The government succeeded in implementing the profit-sharing provisions of the 1917 Constitution, although in the process it had to make some concessions. The organized labor movement would now, for the first time, receive a share of industry's profits, but its share would not be as large as it would have liked. The private sector was satisfied that it had managed to tone down somewhat the redistributive pinch of profit sharing.

This absence of obvious winners and losers can be considered a significant feature of the Mexican decision process. If a political system composed of disparate groups is to remain integrated and viable, all participating groups must be made to feel that they have something to gain from their continued allegiance to the regime. Preventing the creation of easily identifiable winners and losers is thus an important aspect of system maintenance.

To say that there were no obvious winners or losers, however, is not to say that all groups benefited equally. In terms of the concessions they were able to wrest from the government, the business interests can be considered to have gained more than their labor counterparts. Labor, more often than not, was required to pay for the concessions granted by the government to the private sector.

There were several reasons for this outcome. Of primary importance is the differential dependence of business and labor on the government. The labor groups, because they were more dependent on the government, were less able to exert pressure on behalf of their interests than were the relatively more autonomous business groups. The even more dependent Mexican peasants fare still worse. Ronfeldt's recent book on the ejidatarios of Atencingo, for example, vividly illustrates the fruitlessness of the efforts of even rather sophisticated peasant groups to wrest minor concessions from the government.[15]

In addition to showing the impact of differential dependence, the profit-sharing decision also highlights the direct, positive relation between level of mobilization and the ability to force the government to attend to one's demands. The private sector, which was able to win concessions from the labor movement with the support of the government, in general is characterized by a higher level of mobilization than is the labor movement. The announcement of the impending establishment of a compulsory profit-sharing system further mobilized the private sector, since it feared that the decision would seriously damage its economic interests. The normally less mobilized labor movement, on the other hand, was not moved to higher levels of mobilization by the decision, since it regarded the decision as a "gift" from the government, one for which it had not worked particularly hard. Labor therefore did not need to mobilize to protect its interests, a situation that made it more vulnerable to pressures from the government and, indirectly, from the private sector.

The substantial advantages that accrue to the largest and best organized groups within each sector is also apparent from the profit-sharing study. Not all members of the private sector benefited equally from the profit-sharing decision. It is no coincidence that the final formula was one that would least affect the profits of Mexico's largest, most modern, and most highly capitalized firms. These are precisely the firms that dominate CONCAMIN, and they were best represented on the National Profit-Sharing Commission. Smaller, more labor-intensive firms were virtually unrepresented, both within the major business organizations and on the commission. Their exclusion is reflected in the relatively high percentage of their profits that they must distribute among their employees.

A similar situation existed within the labor sector. Labor's representatives on the National Profit-Sharing Commission were the leaders of the country's largest and best organized labor unions and confederations. They dominated the labor movement and were also among the highest paid workers in Mexico. The final profit-sharing formula reflected the strength of their position, for it

143

provided for the distribution of half of the profits to be shared in proportion to the salary of an employee. The highest paid workers, a category that included most union officials and by extension, the labor representatives on the National Commission as well, would thus maintain their economic advantage over their less well-paid colleagues. Furthermore, the profit-sharing system did not apply to Mexico's unorganized rural workers, who constituted the most poverty-stricken segment of the labor force. The rural workers were totally unrepresented on the National Profit-Sharing Commission.

Differences in relative dependence upon the government, group size, organizational strength, and level of mobilization do not, however, entirely explain the outcome of the profit-sharing decision. Despite their substantial advantages over the labor groups, for example, the business interests were unable to pressure the labor organizations into substituting for profit sharing a system of economic bonuses. Even the less drastic proposals of the private sector, like the request for a deduction from profits of a percentage of a firm's total capital investment, came to naught. The explanation for the failure of the private sector to translate its superior strength over labor into concrete economic and political gains lies in the power and commitment of the Mexican government.

The government was not a neutral observer in the profit-sharing decision. Nor was it an uninterested arbiter of conflicting interests. From the very beginning it made clear that it was very interested in the decision and that it was determined to implement the profit-sharing provisions of the 1917 Constitution despite objections or obstruction from the private sector and the absence of an overwhelming desire for profit sharing on the part of the organized labor movement. The profit-sharing study thus indicated the existence of a strong "state interest," which was different from and independent of the interests of any particular group in Mexican society.

The existence of this state interest substantially affected the ability of all involved groups to translate their demands into effective policy in a way that reflected their relative, although limited, power capabilities. In other words, unlike a corresponding decision in a democratic pluralist system, the final decision was not a mirror reflection of the substantially greater autonomy and influence of the private sector, although business did fare somewhat better than it had originally anticipated. Rather, in this particular case, the more highly dependent and less mobilized labor groups gained far more than they could have gained in a democratic pluralist system, basically because the Mexican government was determined to "give" the workers a viable profit-sharing system and to keep the private sector from thwarting its intentions.

A strong state interest in implementing a particular decision in itself does not ensure that the government will obtain its goal. Techniques and procedures that facilitate the translation of the state interest into authoritative decisions are also required. Both elements were present in the case of the profit-sharing decision, and their co-existence accounts for its outcome.

It might be argued that the government did not give the workers a very redistributive profit-sharing system and that this fact reflects the power of the private sector to thwart the supposed state interest. This argument would equate the situation in Mexico with that in a democratic pluralist system where a state interest is often lacking and in its stead one finds only a plethora of private interests, the strongest of which becomes the so-called state or public interest.

The problem with this interpretation of the profit-sharing decision's outcome is that it ignores the possibility of congruence between certain aspects of the state interest and those of particular groups in society. In this view, the existence of a state interest could only be demonstrated if it were radically different from others in the society, say, if it were oriented toward a radical redistribution of income or perhaps even the destruction of the private sector and the substitution of a socialist system for Mexico's present state capitalist economic system.

None of the information gathered during the study of the profit-sharing decision provided evidence that the government had such radical inclinations. Rather, the data point to a strong desire on the part of the López Mateos regime to "do something for labor," in part because of a long period of government neglect and the consequent growing restlessness and dissatisfaction of the rank-and-file membership. The fact that the implementation of the profit-sharing provisions of the revolutionary 1917 Constitution would imbue the decision with great symbolic value was an important consideration. As already noted, the president's decision represented essentially a political commitment to labor, the details of which were to be worked out by government technocrats who gave no indication of being rebels against Mexico's state capitalist economic system. On the contrary, once the president's "decision in principle" had been made, the technocrats perceived their job as upholding the president's commitment while assuring that the final profit-sharing formula would not prove detrimental to Mexico's rather spectacular economic growth record. For this reason, the resulting formula did not deprive the private sector of substantial profits.

In the case of profit sharing, therefore, substantial areas of congruence existed between the state interest and the interests of the private sector. The congruence was not absolute, however, or business would have succeeded in its attempt to eliminate the proposed profit-sharing system and labor would have been at the mercy of the private sector. The compromises and concessions obtained by the private sector were minor in comparison with its inability to obtain its foremost demand, nonimplementation of a compulsory profit-sharing system, or its next demand, a deduction from profits based on a percentage of a firm's total capital investment. This fact attests to the existence of a state interest that encompassed basic areas of disagreement with the private sector.

Based on the evidence from the case study, therefore, the government was the primary beneficiary of the profit-sharing decision. Despite some minor concessions to the private sector, it essentially got what it wanted. It established a viable profit-sharing system that would not threaten Mexico's economic growth potential. It received the gratitude of the labor movement, and, in the end, even the business groups agreed to curtail their opposition and cooperate with the new system. After the government, organized labor received the next greatest share of the benefits of the decision. Despite the concessions extracted by the business groups, labor received a share of industry's profits and the private sector was subjected to an obligatory profit-sharing system.

To determine who benefits most after the government on a more general level is more complicated, however. A number of criteria come into play in any decision. Some of the relatively constant factors are a group's size, organizational strength, level of mobilization, and relative autonomy from the government. Most crucial, however, is a potentially more variable element, the degree of congruence between the state interest on a particular issue and that of any private group. Factors such as a group's size and degree of autonomy are only accurately reflected in a decision's outcome when there is no clearly defined state interest. An example is the situation in the consumer goods industry where, according to Brandenburg, the government has no clear policy. The private sector therefore charges what it wishes, and the majority of all goods purchased on credit by consumers, a group totally lacking in organization, must be returned or repossessed.[16] In those cases in which the state interest is diametrically opposed to that of a particular group, the power capabilities of that group are least accurately reflected in the decision outcome.

In view of the fact that the profit-sharing decision is one of the few redistributive decisions made by the Mexican government in recent years and that it was not radically redistributive, it seems fair to conclude that the government is not yet interested in profoundly altering the status quo. All evidence points to a state interest that is concerned with perpetuating a functioning political and economic system. Minor adjustments may be made, however, such as some slight redistribution in favor of less privileged groups that is deemed necessary to preserve the existing situation. Thus the private sector, whose interests have been most congruent with those of the government during the past few decades, will continue to benefit most consistently from government policies. Next in line will be the middle-sector groups, followed by organized workers, and, finally, organized peasants.

It is conceivable, however, that economic and political conditions may require a more drastic departure from status quo politics in the future. Political discontent over failure to redistribute resources adequately, for example, may result in substantial mass mobilization that could threaten the continued viability of the regime. Should such a situation develop, the evidence from the profit-sharing decision indicates that the government possesses the capabilities to alter its priorities. After all, state interest is not a static concept: its main goal is the preservation of the extant regime. Should circumstances necessitate the redirection of the state interest away from currently favored groups and toward presently neglected ones, there is no obvious reason for believing that the government will be unable to use the authoritarian decision procedures that have been described in this work to convert a redefined state interest into effective policies of a different nature.

Appendix

Interviews

In the course of researching the profit-sharing decision, I interviewed a total of 63 individuals. Most of the interviews occurred between February and June 1968 in Mexico City, after I had exhausted written sources of information. My initial list included all the formal members of the National Profit-Sharing Commission, the leaders of the principal private-sector and labor organizations who were not members of the commission, and other individuals whose names had appeared in the written sources previously examined.

I had been warned that it would be very difficult to gain access to the people with whom I wished to speak since most were important national figures and I was an unknown and not particularly "well-connected" graduate student from the United States. Nevertheless, with one exception, I was able to obtain interviews with everyone whom I contacted. Most of the interviews resulted from a telephone call to the individual in question during which I introduced myself, explained what I was doing in Mexico, and requested an opportunity to discuss the profit-sharing decision.

The interviews ranged from fifteen minutes to three hours in length, depending on the informant's willingness to discuss his role in the decision and the significance of his role. The former was usually the more crucial factor in determining the length of an interview.

All interviews were unstructured and none were taped. I do not know whether people would have been willing to have their remarks taped, but I felt that the use of a tape recorder would have deprived me of much useful "off the record" information. At the beginning of each interview I asked the informant whether I could take notes. Most individuals gave their permission. When an informant preferred that I not take notes, I wrote down what I remembered of the interview immediately after leaving the informant. As a result of these procedures, only informal notes of the interviews exist.

Because I interviewed all but four of the members of the National Profit-Sharing Commission, their names are presented in an order that reflects the organizational structure of the commission. The remaining informants are then listed in alphabetical order.

Members of the National Profit-Sharing Commission

President	Hugo B. Margáin (interviewed in Washington, D.C.)
Technical Director	Octavio A. Hernández

Representatives of Organized Labor

Principal representatives

Adolfo Flores Chapa	The Miners' Union
Alberto Juárez Blancas	The CROC
Jesús Yurén Aguilar	The CTM
Enrique Rangel	The CROC
Blas Chumacero	The CTM

Alternate representatives (possessing voting rights)

Alfredo Rodríguez	The Miners' Union
Francisco Ballina Tabares	The Association of Mexican Pilots
José Ortiz Petriccioli	The CROM
Samuel Ruiz Mora	The FOR
Francisco Benítez	The Theatrical Federation

Representatives of the Private Sector
  Principal representatives
  José Campillo Sainz   CONCAMIN
  Carlos Isoard   CONCAMIN
  Ramiro Alatorre   CONCAMIN
  Heriberto Vidales (not   CONCANACO
    interviewed)
  Ricardo García Sainz   CONCAMIN
  Alternate representatives (possessing voting rights)
  Rafael Lebrija   CONCAMIN
  César Roel   COPARMEX
  Fernando Yllanes Ramos   CONCAMIN
  Genaro García   CONCANACO
  Alfonso Ortega Vélez (not   CONCANACO
    interviewed)
Advisers representing the government
  Carlos Bergés (not interviewed)
  Everardo Gallardo (interviewed by telephone)
Advisers named by organized labor
  Juan Moisés Calleja   The CTM
  Joaquín Gamboa Pascoe   The CTM
Advisers named by the private sector
  Guillermo Prieto Fortún
  Gustavo Romero Kolbeck
Advisers of the technical staff
  Roberto Martínez Le Clainche (interviewed by telephone)
  Salvador Mora Hurtado
Secretaries to the Commission's President and Technical Director
  Luis Riba Jr.
  Jaime H. Castellanos
Technical Staff
  Alfredo Uruchurtu
  Emilio Carrillo
  Enrique Araujo
  Manuel Rodríguez Rocha
  Ernesto Valderrama Herrera
  Manuel Torres M.
Head of the Administrative Department
  Raúl Adame Bahena (not interviewed)

Other Informants

Alfredo Balmaceda Estrada
  Head of the Profit-Sharing Department of the Ministry of Labor
James Creagan
  Labor attaché, United States Embassy, Mexico City
Mario de la Cueva
  Member of the commission that drafted the 1961 constitutional
  amendment and the 1962 labor law reforms
Vicente Fuentes Díaz
  PRI official
Salomón González Blanco
  Secretary of labor under President López Mateos
Atanasio González Martínez
  Head of the Profit-Sharing Department of the Treasury
Roberto Guajardo Suárez
  President of COPARMEX
Román Iglesias González
  Head of the Profit-Sharing Department of the Income Tax
  Division of the Treasury (in September 1972, when interviewed)
Murray Kaufman
  Business attaché of the United States Embassy, Mexico City
Juan Landerreche Obregón
  PAN leader
Francisco Lerdo de Tejada
  COPARMEX official
Carlos Madrazo
  Former president of the PRI
Rodolfo Martínez López
  COPARMEX economist (interviewed in September 1972)
Gerardo Medina Valdéz
  PAN congressman
Robert Ozanne
  Professor of economics, University of Wisconsin
Francisco Parra
  AMIS functionary
Agustín Reyes Ponce
  Labor law authority employed by CONCAMIN
Enrique Ramírez y Ramírez
  Editor of *El Día*

José Luis Robles Glenn
  Lawyer for the private sector
Rosendo Salazar
  Former CTM leader
Irving Salert
  Labor attaché, United States Embassy, Mexico City
María Salmorán de Tamayo
  Member of the commission that drafted the 1961 constitutional
  amendment and the 1962 labor law reforms
Juan Sánchez Navarro
  President of CONCAMIN
Robert E. Service
  Vice-consul, United States Embassy, Mexico City
Alberto Trueba Urbina
  National authority on Mexican labor law
Victor Urquidi
  Professor of economics and head of El Colegio de México
Fidel Velázquez
  Secretary General of the CTM
Al Wichtrich
  President of the U.S. Chamber of Commerce in Mexico
Gerardo Zavala
  Head of the Profit-Sharing Department of the Ministry of Labor
  (in September 1972, when interviewed)

Notes

Chapter 1: Introduction

1. This argument is made most strongly by Robert E. Scott in his *Mexican Government in Transition* (Urbana: University of Illinois Press, 1959). Other writers who stress the importance of the PRI in the decision-making process, although to a somewhat lesser degree, are L. Vincent Padgett, *The Mexican Political System* (Boston: Houghton-Mifflin Company, 1966); and Raymond Vernon, *The Dilemma of Mexico's Development* (Cambridge, Mass.: Harvard University Press, 1963).

2. The earliest of these works is Frank Brandenburg's *The Making of Modern Mexico* (Englewood Cliffs, N.J.: Prentice-Hall, 1964). See also Roger D. Hansen, *The Politics of Mexican Development* (Baltimore, The Johns Hopkins Press, 1971); and Robert E. Scott, "Mexico: The Established Revolution," in *Political Culture and Political Development*, ed. Lucien Pye and Sidney Verba (Princeton: Princeton University Press, 1965), pp. 330-395. For a more detailed discussion of the contrasting viewpoints, see Carolyn and Martin Needleman, "Who Rules Mexico? A Critique of Some Current Views of the Mexican Political Process," *Journal of Politics*, 31 (November 1969): 1011-1034.

3. See especially Vernon, *The Dilemma*, and Scott, *Mexican Government*.

4. For this argument, the most relevant works are Hansen, *Politics of Mexican Development*; Brandenburg, *Making of Modern Mexico*, and Charles W. Anderson, "Bankers as Revolutionaries: Politics and Development Banking in Mexico," in *The Political Economy of Mexico*, ed. William P. Glade, Jr., and Charles W. Anderson (Madison: University of Wisconsin Press, 1963).

5. See particularly Pablo González Casanova's argument regarding the "marginality" of most Mexicans in *La democracia en México* (México: Ediciones Era, 1967), and Guillermo A. O'Donnell's characterization of Mexico as an exclusionary bureaucratic-authoritarian regime in *Modernization and Bureaucratic-Authoritarianism: Studies in South American Politics* (Berkeley: Institute of International Studies, University of California, 1973).

6. This argument is made most forcefully by Hansen, *Politics of Mexican Development*, and Clark W. Reynolds, *The Mexican Economy: Twentieth-Century Structure and Growth* (New Haven: Yale University Press, 1970).

7. See especially González Casanova, *La democracia en México*; Martin C. Needler, *Politics and Society in Mexico* (Albuquerque: University of New Mexico Press, 1971); Scott, *Mexican Government*; and John Walton and Joyce A. Sween, "Urbanization, Industrialization and Voting in Mexico: A Longitudinal Analysis of Official and Opposition Party Support," *Social Science Quarterly 52* (December 1971): 721-745.

8. The principal exponent of this view is Vernon in *The Dilemma*.

9. See especially Hansen, *Politics of Mexican Development*; O'Donnell, *Modernization and Bureaucratic-Authoritarianism*; and Susan Kaufman Purcell, "Decision-Making in an Authoritarian Regime: Theoretical Implications from a Mexican Case Study," *World Politics 26* (October 1973): 28-54.

10. My description of the characteristics of an authoritarian regime is drawn from the following works: Juan J. Linz, "An Authoritarian Regime: Spain" in *Cleavages, Ideologies and Party Systems: Contributions to Comparative Political Sociology*, ed. E. Allardt and Y. Littunen (Helsinki: Transactions of the Westermarck Society, 1964), pp. 291-341; Susan Kaufman Purcell, "Authoritarianism," *Comparative Politics 5* (January 1973): 301-312; Purcell, "Decision-Making in an Authoritarian Regime"; Philippe C. Schmitter, *Interest Conflict and Political Change in Brazil* (Stanford: Stanford University Press, 1971); Philippe C. Schmitter, "Paths to Political Development in Latin America," in "Changing Latin America: New Interpretations of Its Politics and Society," ed. Douglas A. Chalmers, *Proceedings of the Academy of Political Science 30* (August 1972): 83-108; Howard J. Wiarda, "Toward a Framework for the Study

of Political Change in the Iberic-Latin Tradition: The Corporative Model," *World Politics* 25 (January 1973): 206-235; O'Donnell, *Modernization and Bureaucratic-Authoritarianism;* Charles W. Anderson, *The Political Economy of Modern Spain: Policy-Making in an Authoritarian System,* (Madison, University of Wisconsin Press, 1970); Juan J. Linz, "Notes toward a Typology of Authoritarian Regimes" (Paper prepared for delivery at the 1972 Annual Meeting of the American Political Science Association, Washington, D.C.), mimeographed; Douglas A. Chalmers, "Political Groups and Authority in Brazil: Some Continuities in a Decade of Confusion and Change," in *Brazil in the Sixties,* ed. Riordan Roett (Nashville: Vanderbilt University Press, 1972), pp. 51-76; Kalman A. Silvert, "The Costs of Anti-Nationalism: Argentina," (for the related concept of "Mediterranean syndicalism") in *Expectant Peoples,* ed. K. H. Silvert (New York: Vintage Books, 1963), pp. 347-371; Ronald Rogowski and Lois Wasserspring, *Does Political Development Exist? Corporatism in Old and New Societies,* Sage Professional Papers in Comparative Politics (Beverly Hills, Calif.: Sage Publications, 1971); Robert R. Kaufman, "Corporatism, Clientelism, and Partisan Conflict in Latin America: A Comparative Study of Argentina, Brazil, Chile, Colombia, Mexico, Venezuela, and Uruguay" (Unpublished manuscript, 1973); James M. Malloy, "Authoritarianism, Corporatism and Mobilization in Peru," *The Review of Politics* 36 (January 1974): 52-84; and Philippe C. Schmitter, "Still the Century of Corporatism?" *The Review of Politics* 36 (January 1974): 85-131.

11. For the concept of "limited pluralism" see Linz, "An Authoritarian Regime: Spain," and Purcell, "Decision-Making in an Authoritarian Regime." For "corporate pluralism" see Schmitter, *Interest Conflict;* Samuel P. Huntington, "Social and Institutional Dynamics of One-Party Systems," in *Authoritarian Politics in Modern Society: The Dynamics of Established One-Party Systems,* ed. Samuel P. Huntington and Clement H. Moore (New York: Basic Books, 1970), p. 35; and Andrew C. Janos, "Group Politics in Communist Society: A Second Look at the Pluralistic Model," in Huntington and Moore, *Authoritarian Politics,* p. 441. The concept "ordered pluralism" is from Jane S. Jaquette, "Revolution by Fiat: The Context of Policy Making in Peru," *The Western Political Quarterly* 25 (December 1972): 649.

12. Linz, "An Authoritarian Regime: Spain," p. 304.

13. Linz, "Notes toward a Typology," p. 27.

14. O'Donnell, *Modernization and Bureaucratic-Authoritarianism,* p. 112.

15. Ibid.

16. Linz, "Notes toward a Typology," p. 28.

17. Schmitter, *Interest Conflict*, p. 378.

18. Rogowski and Wasserspring, *Does Political Development Exist?* p. 31.

19. Kaufman, "Corporatism, Clientelism, and Partisan Conflict."

20. Samuel P. Huntington, *Political Order in Changing Societies,* (New Haven: Yale University Press, 1970), p. 80.

21. Ibid., p. 82.

22. O'Donnell, *Modernization and Bureaucratic-Authoritarianism,* pp. 95-97.

23. For additional information regarding the interviews, see the appendix.

Chapter 2: The Political Environment

1. For a more detailed description of the regime of Porfirio Díaz, see Daniel Cosío Villegas et al., *Historia moderna de México,* 7 vols. (México, D.F.: Editorial Hermes, 1955-1957), and José C. Valadés, *El porfirismo: Historia de un régimen* (México, D.F.: Antigua Librería Robledo, de José Porrúa e hijos, 1941).

2. Howard F. Cline, *Mexico—Revolution to Evolution, 1940-1960* (London: Oxford University Press, 1962), pp. 142-143.

3. Ibid.

4. Robert E. Scott, "Legislatures and Legislation," in *Government and Politics in Latin America,* ed. Harold Eugene Davis (New York: The Ronald Press, 1958), p. 169.

5. Howard F. Cline, *The United States and Mexico* (New York: Atheneum, 1963), p. 169.

6. Frank Tannenbaum, *Mexico: The Struggle for Peace and Bread* (New York: Alfred A. Knopf, 1950), p. 118. For a recent discussion of corporatist aspects of Mexican politics, see Ronald Rogowski and Lois Wasserspring, *Does Political Development Exist? Corporatism in Old and New Societies,* Sage Professional Papers in Comparative Politics (Beverly Hills, Calif.: Sage Publications, 1971).

7. For a discussion of the corporate features of the colonial system, see Magali Sarfatti, *Spanish Bureaucratic-Patrimonialism in America* (Berkeley: Institute of International Studies, University of California, 1966), pp. 16-17.

8. Robert E. Scott, *Mexican Government in Transition* (Urbana: University of Illinois Press, 1959), p. 101.

9. The term "constitutionalism" is from Frank Brandenburg, *The Making of Modern Mexico* (Englewood Cliffs, N.J.: Prentice-Hall, 1964), p. 10.

10. Ibid., pp. 10-11.

11. Patricia McIntire Richmond, "Mexico: A Case Study of One-Party Politics" (Ph.D. diss., University of California, Berkeley, 1965), pp. 190-191; Linda Mirin and Arthur L. Stinchcombe, "The Political Mobilization of Mexican Peasants," mimeographed (Baltimore: Johns Hopkins University), p. 21; and Robert Frank Adie, "Agrarianism in the Mexican Political System" (Ph.D. diss., University of Texas, Austin, 1970), pp. 135-139. See also William S. Tuohy and David Ronfeldt, "Political Control and the Recruitment of Middle-Level Elites in Mexico: An Example from Agrarian Politics," *Western Political Quarterly* 22 (June 1969): 365-374, for a case study of a CNC "election."

12. For a discussion of the CCI see Bo Anderson and James D. Cockcroft, "Control and Cooptation in Mexican Politics," in *Latin American Radicalism: A Documentary Report on Left and Nationalist Movements*, ed. Irving Louis Horowitz, Josué de Castro, and John Gerassi (New York: Random House, 1969), pp. 366-389.

13. Richmond, "Mexico," pp. 227-228.

14. Roger D. Hansen, *The Politics of Mexican Development* (Baltimore: The Johns Hopkins Press, 1971), pp. 116-120; and Alonso Aguilar M. and Fernando Carmona, *México: Riqueza y miseria* (México, D.F.: Editorial Nuestro Tiempo, 1967), pp. 21-24.

For an excellent case study of the way in which the dependent situation of the ejidatarios frustrates their attempts to improve their status see David Ronfeldt, *Atencingo: The Politics of Agrarian Struggle in a Mexican Ejido* (Stanford: Stanford University Press, 1973).

15. Richard Ulric Miller, "The Role of Labor Organizations in a Developing Country: The Case of Mexico" (Ph.D. diss. Cornell University, 1966), pp. 63-65.

16. For additional information on the history of the organized labor movement in Mexico see Luis Arraiza, *Historia del movimiento obrero mexicano* (México, D.F.: Editorial Cuahtémoc, 1964); Joe C. Ashby, *Organized Labor and the Mexican Revolution under Lázaro Cárdenas* (Chapel Hill: University of North Carolina Press, 1967); Marjorie Ruth Clark, *Organized Labor in Mexico* (Chapel Hill: University of North Carolina Press, 1934); Robert Paul Millon, *Mexican Marxist—Vicente Lombardo Toledano* (Chapel Hill: University of North Carolina Press, 1966); Alfonso López Aparicio, *El movimiento obrero en México* (México, D.F.: Editorial Jus, 1958); and Vicente Lombardo Toledano, *Teoría y práctica del movimiento sindical mexicano* (México, D.F.: Editorial del Magisterio, 1961).

17. Miller, "The Role of Labor Organizations," pp. 70, 292.

18. Michael David Everett, "The Role of the Mexican Trade Unions, 1950-1963" (Ph.D. diss., Washington University, Missouri, 1967), p. 22.

19. Ibid., p. 90, and Millon, *Mexican Marxist*, p. 143.

20. Miller, "The Role of Labor Organizations," p. 80.

21. Ibid., p. 293.

22. Everett, "Role of the Mexican Trade Unions," p. 71. Other conditions required for a strike to be declared existent are: that the strike have as its goal "to obtain from an employer the conclusion of or compliance with a collective contract of employment," "to exact a revision of a collective contract if necessary, upon the expiration of its period of operation, under conditions and in the cases laid down in this act," "to support a strike which has for its object any of the purposes enumerated in the preceding items and which has not been declared illegal" (Miller, "The Role of Labor Organizations,'" p. 81).

23. Everett, "Role of the Mexican Trade Unions," p. 18. Sometimes dues are considerably lower. Miller reports that CTM dues average 4 pesos (U.S. $0.32) from each affiliate. Miller, "The Role of Labor Organizations," p. 51.

24. A 1959 report of the national committee of the CTM, for example, complained that only 20 percent of the confederation's member unions were paid up for that year. Miller, "The Role of Labor Organizations." See also Antonio Ugalde, *Power and Conflict in a Mexican Community* (Albuquerque: University of New Mexico Press, 1970), p. 26.

25. Interview with a labor leader, 1967, Mexico City.

26. Simon Rottenberg, "México: Trabajo y desarrollo económico," *Foro Internacional* 11 (July-September 1959): 98.

27. Miller, "The Role of Labor Organizations," p. 71.

28. Everett, "Role of the Mexican Trade Unions," p. 25.

29. *La Prensa*, April 30, 1968, p. 2; and Article 49 of the Federal Labor Law.

30. Most of the national unions were established before 1940, principally during the Cárdenas regime as a result of his encouragement. Since then, neither the government nor the national leaders of the labor movement have displayed any interest in increasing the number of such unions. Miller, "The Role of Labor Organizations," p. 54.

31. Lombardo Toledano, *Teoría y práctica*, p. 88.

32. Scott, *Mexican Government in Transition*, pp. 165-168; Everett, "Role of the Mexican Trade Unions," pp. 93-94; and Miller, "The Role of Labor Organizations," pp. 39-40.

33. Scott, *Mexican Government in Transition*, p. 165.

34. Ugalde, *Power and Conflict*. Richard R. Fagen and William S. Tuohy also note the competition between the CTM and the CROC in Jalapa in *Politics and Privilege in a Mexican City* (Stanford: Stanford University Press, 1972), p. 59.

35. Clark, *Organized Labor in Mexico*, chap. 4.

36. Victor Alba, *Historia del movimiento obrero en América Latina*, (México, D.F.: Libreros Mexicanos Unidos, 1964), p. 449.

37. Scott, *Mexican Government in Transition*, p. 164.

38. Robert J. Alexander, *El movimiento obrero en América Latina*, (México, D.F.: Editorial Roble, 1967), p. 249.

39. Pablo González Casanova, *La democracia en México* (México, Ediciones Era, 1965), p. 24. For a similar conclusion regarding the positive correlation between the number of strikes and the favorable attitude of the president toward organized labor see Vicente Fuentes Díaz, "Desarrollo y evolución del movimiento obrero a partir de 1929," *Revista de Ciencias Políticas y Sociales* 17 (July-September 1959): 325-348; and Guadalupe Rivera Marín, "Los conflictos de trabajo en México, 1937-1950," *El Trimestre Económico* 22 (April-June 1955): pp. 181-208; and James W. Wilkie, *The Mexican Revolution: Federal Expenditure and Social Change since 1910* (Berkeley: University of California Press, 1967), pp. 187-189.

40. Raymond Vernon, *The Dilemma of Mexico's Development* (Cambridge, Mass.: Harvard University Press, 1963), p. 101.

41. González Casanova, *La democracia en México*, pp. 183-184.

42. Fuentes Díaz, "Desarrollo y evolución," pp. 345-346.

43. Rivera Marín, "Los conflictos de trabajo," p. 203.

44. John Isbister, "Urban Employment and Wages in a Developing Economy: The Case of Mexico," *Economic Development and Cultural Change* 20 (October 1971): 35-39. See also Wilkie, *The Mexican Revolution*, p. 187.

45. The term "middle sectors" is John J. Johnson's. See his *Political Change in Latin America: The Emergence of the Middle Sectors* (Stanford: Stanford University Press, 1958).

46. David Schers, "The Popular Sector of the Mexican PRI" (Ph.D. diss., University of New Mexico, 1972), pp. 203-204.

47. Ibid., pp. 106, 102.

48. Ibid., p. 181.

49. Clark W. Reynolds, *The Mexican Economy: Twentieth-Century Structure and Growth* (New Haven: Yale University Press, 1970), p. 308.

50. Hansen, *Politics of Mexican Development*, p. 5. Between 1900 and 1910, approximately two-thirds of all investment was foreign in origin. Since 1940, however, domestic savings have financed approximately 90 percent of all gross fixed investment. Ibid., p. 42.

51. Ibid., p. 56.

52. For more detailed information about the organizations of the private sector see Robert Jones Shafer, *Mexican Business Organizations: History and Analysis* (Syracuse: Syracuse University Press, 1973); Frank R. Brandenburg, "Organized Business in Mexico," *Inter-American Economic Affairs* 12 (Winter 1958): 26-50; Brandenburg, *The Making of*

NOTES

Modern Mexico; Isaac Guzmán Valdivia, "El movimiento patronal," in *México: 50 años de revolución,* ed. Julio Durán Ochoa et al., vol. 2, *La vida social* (México: Fondo de Cultura Económica, 1961); and Vernon, *The Dilemma of Mexico's Development.*

53. Vernon, *The Dilemma of Mexico's Development,* pp. 166-172.

54. COPARMEX is an exception. It has had the same leader for approximately two decades.

55. Brandenburg, *The Making of Modern Mexico,* p. 230. The Mexican government also owns or controls other industries that, while not economically important, are politically significant. For example, the manufacture and distribution of newsprint and the distribution and exhibition of motion pictures are under government control.

56. Frank Brandenburg, "The Relevance of Mexican Experience to Latin American Development," *Orbis 9* (Spring 1965): 196.

57. Hansen, *Politics of Mexican Development,* p. 44.

58. Enrique Pérez López, "The National Product of Mexico: 1895 to 1964," in Enrique Pérez López et al., *Mexico's Recent Economic Growth: The Mexican View* (Austin: University of Texas, Institute of Latin American Studies, 1967), p. 33.

59. Shafer, *Mexican Business Organizations,* p. 8. For an excellent analysis of the role of Mexican development banks see Charles W. Anderson, "Bankers as Revolutionaries: Politics and Development Banking in Mexico," in *The Political Economy of Mexico,* ed. William P. Glade, Jr., and Charles W. Anderson (Madison: University of Wisconsin Press, 1968), pp. 103-191.

60. Brandenburg, *The Making of Modern Mexico,* p. 231.

61. Reynolds, *The Mexican Economy,* p. 255.

62. Ibid., p. 287; and John E. Koehler, *Economic Policy-Making with Limited Information: The Process of Macro-Control in Mexico* (Santa Monica, Calif.: The Rand Corporation, August 1968), p. 58.

63. Reynolds, *The Mexican Economy,* pp. 284-289. In 1962, investment of government and government enterprise as a share of gross investment equalled 32.7 percent, 10 percent more than in 1956.

64. Vernon, *The Dilemma of Mexico's Development,* p. 26.

65. Reynolds, *The Mexican Economy,* p. 186.

66. Hansen, *Politics of Mexican Development,* p. 48.

67. Ibid., p. 49. See also Reynolds, *The Mexican Economy,* p. 190; and Bernard S. Katz, "Mexican Fiscal and Subsidy Incentives for Industrial Development," *The American Journal of Economics and Sociology* 31 (October 1972): 353-358.

68. Hansen, *Politics of Mexican Development,* p. 45. For a more detailed analysis of business-government relations in Mexico see John F. H. Purcell and Susan Kaufman Purcell, "Mexican Business and Public Policy"

(Paper presented at a conference on Authoritarianism and Corporatism in Latin Ameria, University of Pittsburgh, April 4-6, 1974).

69. In 1969, for example, opposition parties held only 17 of the more than 2,300 mayoralities. Of the 700 most important municipal councils, opposition parties held majorities on only 23. Hansen, *Politics of Mexican Development*, pp. 102-103.

70. All figures are from James W. Wilkie, "New Hypotheses for Statistical Research in Recent Mexican History," *Latin American Research Review* 6 (Summer 1971): 5.

71. Wilkie, *The Mexican Revolution* p. 182. The 1970 figure is from Ronald H. McDonald, *Party Systems and Elections in Latin America* (Chicago: Markham Publishing Co., 1971), p. 253.

72. John Walton and Joyce A. Sween, "Urbanization, Industrialization and Voting in Mexico: A Longitudinal Analysis of Official and Opposition Party Support," *Social Science Quarterly* 52 (December 1971): 735, 741; Robert K. Furtak, "El Partido Revolucionario Institucional: Integración nacional y movilización electoral," *Foro Internacional* 9 (April-June 1969): 349-350; José Luis Reyna, "Desarrollo económico, distribución del poder y participación política: El caso mexicano," *Ciencias Políticas y Sociales* 13 (October-December 1967): 469-486. Ugalde makes the same point with regard to Ensenada. See Ugalde, *Power and Conflict*, p. 153.

73. Furtak, "El Partido Revolucionario Institucional," p. 350; Barry Ames, "Bases of Support for Mexico's Dominant Party," *The American Political Science Review* 64 (March 1970): 167; and Walton and Sween, "Urbanization, Industrialization, and Voting," p. 735.

74. González Casanova, *La democracia en México*, p. 117. Neither the level of turnout nor the direction of the vote is related to party *affiliation*, however. Ames, "Bases of Support for Mexico's Dominant Party," p. 167; and Furtak, "El Partido Revolucionario Institucional," p. 350.

75. González Casanova, *La democracia en México*, p. 27. Approximately 90 percent of the legislation passed by Congress originates with the executive. Estados Unidos Mexicanos, *Por el camino de un México nuevo-Origen, significado y perspectivas de la Constitución de 1917*, Edición de la XLVI Legislatura de la Cámara de Diputados (Mexico, D.F., 1967), p. 204.

76. Despite the numerous constraints upon independent action by congressmen, members of the Chamber of Deputies in particular have managed occasionally to pass legislation that the president did not wish to have become law. In such cases, the Senate, whose ranks are filled with older individuals whose loyalty to the system is more tested, refused to act on the proposed legislation and thus effectively killed it.

77. My tally, from information in the *Diario de los debates de la Cámara de Diputados.* Despite the existence of minor-party representatives, the single-party mentality is evident from the fact that the *Diario de los debates* does not identify deputies by their party affiliation, nor are votes broken down by party. When I mentioned this fact to one of the archivists of the Chamber of Deputies, he replied, "It doesn't matter. We know which ones are PRI and which ones are not."

78. L. Vincent Padgett, *The Mexican Political System* (Boston: Houghton-Mifflin Co., 1966), p. 159.

79. The 1964 constitutional amendment increasing minority-party representation expanded the possibilities for the introduction of legislative proposals embodying new ideas. Between 1957 and 1961, for example, the small number of PAN deputies introduced only three significant pieces of legislation. (Two of them were not acted upon and the third was defeated.) Between 1964 and 1967, when the PAN was represented by twenty deputies, thirty-four pieces of legislation were introduced. I obtained this information through a survey of the *Diario de los debates de la Cámara de Diputados.*

80. Ashby, *Organized Labor,* p. 74.

81. Richmond, "Mexico," p. 227.

82. Ashby, *Organized Labor,* p. 43.

83. Aguilar and Carmona, *México: Riqueza y miseria,* p. 127.

84. Alba, *Historia del movimiento obrero,* p. 450.

85. "El PRI," *Mañana* 978 (May 26, 1962): 43.

86. The results of a 1963 survey in the Federal District basically support the estimate that 23 percent of the total population belongs to the PRI. In the survey, approximately 80 percent of the respondents reported that they were not affiliated with any political party. *Mañana,* October 5, 1963, quoted in Richmond, "Mexico," p. 320.

87. In 1962, immediately prior to the membership campaign, party members equaled 14 percent of the total population. At the end of 1963, the head of the PRI announced that the party had approximately 7 million members. The following year, the new PRI head claimed 8.6 million members. Richmond, "Mexico," p. 320. In 1967, the PRI was reported to have 8.1 million members. Mario Ezcurdia, *Análisis teórico del Partido Revolucionario Institucional* (México, B. Costa-Amic, 1968), p. 78. In 1962, prior to the membership campaign, the PRI reported 5 million members. González Casanova, *La democracia en México,* p. 255. According to the 1960 census, Mexico's total population is approximately 35 million.

88. Article 10 of the PRI's statutes states that "affiliation with the Partido Revolucionario Institucional will be on an individual basis, whether the application be made directly by the interested party or by the

organized group of the sector to which he belongs . . ." (Partido Revolucionario Institucional, *Estatutos*, 1966, p. 7).

89. Richmond, "Mexico," p. 321.

90. Ibid., pp. 25-26.

91. See, for example, González Casanova, *La democracia en México;* and Guillermo O'Donnell, *Modernization and Bureaucratic-Authoritarianism: Studies in South American Politics* (Berkeley, Institute of International Studies, University of California, 1973). The term "marginal" is from González Casanova, *La democracia en México,* p. 72. The marginal population is essentially the equivalent of Almond and Verba's "parochials." Gabriel A. Almond and Sidney Verba, *The Civic Culture* (Boston: Little, Brown and Co., 1965).

92. Martin C. Needler, *Politics and Society in Mexico* (Albuquerque: University of New Mexico Press, 1971), p. 96.

93. According to a different source, 56.8 percent of the population in the United States over twenty years of age votes, and in Mexico, 49.8 percent of the population over twenty years of age votes. Charles Lewis Taylor and Michael C. Hudson, eds., *World Handbook of Political and Social Indicators,* 2d ed. (New Haven: Yale University Press, 1972), p. 56. Taylor and Hudson rank countries in terms of their voter turnout as a percentage of the population *eligible to vote.* On this list, Mexico ranks 93 out of 104 countries. The list is a poor indicator of mobilization levels, however, because underdeveloped countries that deny the vote to illiterates rank very high. Paraguay, for example, ranks 36 on Taylor and Hudson's list of 104 countries. The United States and Mexico rank 92 and 93 respectively. Unfortunately, Taylor and Hudson do *not* rank countries according to voter turnout as a percentage of the population over twenty years of age.

94. Robert C. Fried, "Mexico City," in *Great Cities of the World: Their Government, Politics and Planning,* ed. William A. Robson (Beverly Hills, Calif.: Sage Publications, 1967) p. 197.

95. Ames, "Bases of Support," p. 165.

96. Wayne A. Cornelius, "Urbanization as an Agent of Latin American Political Instability: The Case of Mexico," *The American Political Science Review* 63 (September 1969): p. 853.

97. Douglas A. Chalmers, "Political Groups and Authority in Brazil: Some Continuities in a Decade of Confusion and Change," in *Brazil in the Sixties,* ed. Riordan Roett (Nashville: Vanderbilt University Press, 1972), pp. 51-76.

98. Ibid., p. 56.

99. González Casanova, *La democracia en México,* pp. 39-41.

100. George Foster, *Tzintzuntzan: Mexican Peasants in a Changing World* (Boston: Little, Brown and Co., 1967), p. 214.

101. The resemblance between the contemporary "style of rulership" and that which existed when Mexico was ruled by the King of Spain and his viceroys has been noted by Woodrow Borah: "A number of us who have seen peasant delegations waiting for the President of Mexico in the presidential patio of the National Palace or in the antechambers of Los Pinos [the presidential residence] have been struck by the fact that we were watching the General Indian Court of New Spain functioning today rather much as it must have when Antonio de Mendoza gave it informal existence or Luis de Velasco II gave it formal structure" (Woodrow Borah, "Colonial Institutions and Contemporary Latin America: Political and Economic Life," *Readings in Latin American History*, ed. Lewis Hanke [New York: Thomas Y. Crowell Co., 1966], 2:21).

102. This use of the term "broker" is from Eric R. Wolf, "Aspects of Group Relations in a Complex Society: Mexico," in *Contemporary Cultures and Societies of Latin America*, ed. Dwight B. Heath and Richard N. Adams (New York: Random House, 1965), pp. 85-101. The term "palanca" is from George M. Foster, "The Dyadic Contract: A Model for the Social Structure of a Mexican Peasant Village," in *Peasant Society: A Reader*, ed. Jack M. Potter, May N. Díaz, George M. Foster (Boston: Little, Brown and Co., 1967), p. 229. All other terms are from González Casanova, *La democracia en México*, pp. 119-120.

103. William D'Antonio and William H. Form, *Influentials in Two Border Cities: A Study in Community Decision-Making* (Southbend, Ind.: University of Notre Dame Press, 1965), p. 172.

104. Ronfeldt, *Atencingo*.

105. Manuel Ávila, *Tradition and Growth: A Study of Four Mexican Villages* (Chicago: University of Chicago Press, 1969), p. 89.

106. Wolf, "Group Relations in a Complex Society," p. 93.

107. González Casanova, *La democracia en México*, p. 121.

108. Max Weber, *The Theory of Social and Economic Organization*, ed. Talcott Parsons (New York: The Free Press, 1964), p. 347. For examples of such delegation of power in Mexico, see Francine F. Rabinovitz, "Decision-Making for Development in Mexico City," mimeographed.

109. Wendell Karl Gordon Schaeffer, quoted in Linda S. Mirin, "Public Investment in Aguascalientes: A Study in the Politics of Economic Policy" (Ph.D. diss., Harvard University, 1964), p. 27.

110. There are, of course, exceptions to this rule. The most notable are Antonio Ortiz Mena, who served as minister of finance under Presidents López Mateos and Díaz Ordaz, and Lic. Ernesto Uruchurtu, who was governor of the Federal District under three different presidents. Uruchurtu had managed to appropriate so much power that it was necessary to mobilize the barrio dwellers against him in order to have an excuse for removing him from office.

111. The Spanish colonial system apparently avoided appropriation of decision-making power by subordinates to whom power was delegated as a result of similar arrangements, as this analysis by John Phelan suggests: "Historians have assumed that the Spanish bureaucracy like other bureaucratic organizations had only one goal or a set of commensurate goals and that standards of conduct for members were not mutually conflicting. If this assumption is cast aside in favor of goal ambiguity and conflicting standards, new light is thrown on the chasm between the law and its observance in the Spanish empire. The wide gap between the two was not a flaw, as had been traditionally assumed. On the contrary, the distance between observance and nonobservance was a necessary component of the system. Given the ambiguity of the goals and the conflict among the standards, all the laws could not be enforced simultaneously. The very conflict among the standards, which prevented a subordinate from meeting all the standards at once, gave subordinates a voice in decision-making without jeopardizing the control of their superiors over the whole system" (John Phelan, "Authority and Flexibility in the Spanish Imperial Bureaucracy," *Administrative Science Quarterly* 5 [1960]: 63-64, cited in Charles Gibson, *Spain in America* [New York: Harper & Row, 1966], p. 110.

112. For the notion of scarcity as a cause of patron-client relations see Sidney Tarrow, *Peasant Communism in Southern Italy* (New Haven: Yale University Press, 1967), pp. 75-76; and Eric R. Wolf, "Types of Latin American Peasantry: A Preliminary Discussion," *American Anthropologist* 3, pt. 1 (June 1955): 465.

113. For a detailed discussion of mobilization and subsequent demobilization efforts under Cárdenas see Wayne A. Cornelius, "Nation Building, Participation and Distribution: The Politics of Social Reform Under Cárdenas," in *Crisis, Choice and Change: Historical Studies of Political Development*, ed. Gabriel A. Almond, Scott C. Flanagan, and Robert J. Mundt (Boston: Little, Brown and Co., 1973), pp. 392-498.

114. The following discussion is largely drawn from John F. H. Purcell and Susan Kaufman Purcell, "Machine Politics and Socio-Economic Change in Mexico," in *Contemporary Mexico: Papers of the IV International Congress of Mexican History*, ed. James W. Wilkie, Michael C. Meyer, and Edna Monzón de Wilkie (Berkeley, Los Angeles, London: University of California Press, 1975).

115. Wayne A. Cornelius, "The Impact of Governmental Performance on Political Attitudes and Behavior: The Case of the Urban Poor in Mexico City," *Latin American Urban Research*, vol. 3, ed. Francine F. Rabinovitz and Felicity M. Trueblood (Beverly Hills, Calif.: Sage Publications, 1973); Ugalde, *Power and Conflict;* Susan Eva Eckstein, "The Poverty of Revolution" (Ph.D. diss. Columbia University, 1972); Antonio Ugalde, Leslie Olson, David Schers, and Miguel Von Hoegen, "The Urbanization

Process of a Poor Mexican Neighborhood: The Case of San Felipe del Real Adicional, Juárez," (1973), mimeographed; and Martin C. Needler, "Political Aspects of Urbanization in Mexico" in *City and Country in the Third World: Issues in the Modernization of Latin America,* ed. Arthur J. Field (Cambridge, Mass.: Schenkman Publishing Co., Inc., 1970) pp. 287-299.

116. See, for example, Ronfeldt, *Atencingo;* William S. Tuohy, "Institutionalized Revolution in a Mexican City" (Ph.D. diss., Stanford University, 1967), p. 69; Foster, *Tzintzuntzan,* p. 100; D'Antonio and Form, *Influentials in Two Border Cities,* p. 172; and Ugalde, *Power and Conflict,* pp. 160-172. For a recent review of Mexican community studies see Susan Kaufman Purcell and John F. H. Purcell, "Community Power and Benefits from the Nation: The Case of Mexico," *Latin American Urban Research,* Vol. 3, ed. Francine F. Rabinovitz and Felicity M. Trueblood (Beverly Hills, Calif.: Sage Publications, 1973), pp. 49-76.

117. D'Antonio and Form, *Influentials in Two Border Cities,* p. 82; Ugalde, *Power and Conflict,* pp. 160-172.

118. Brandenburg estimates that "every six-year administration witnesses a turnover of approximately 18,000 elective offices and 25,000 appointive posts" (Brandenburg, *The Making of Modern Mexico,* p. 157).

119. See also Oscar Lewis, *Life in a Mexican Village: Tepoztlán Restudied* (Urbana: University of Illinois Press, 1951), p. 251; and Ugalde et al., "The Urbanization Process," p. 58.

120. Ezcurdia, *Análisis teórico,* p. 86.

121. See, for example, Ugalde, *Power and Conflict,* p. 139; Lewis, *Life in a Mexican Village,* p. 251; Charles Irving Mundale, "Local Politics, Integration and National Stability in Mexico" (Ph.D. diss., University of Minnesota, 1971), p. 73; Cynthia Nelson, "The Waiting Village: Social Change in a Mexican Peasant Community" (Ph.D. diss., University of California, Berkeley, 1963. Revised and published as *The Waiting Village: Social Change in Rural Mexico* (Boston: Little, Brown and Co., 1971).

122. Fagen and Tuohy, *Politics and Privilege,* p. 62.

123. Mundale, "Local Politics, Integration and National Stability," pp. 100-101.

124. William P. Tucker, *The Mexican Government Today* (Minneapolis: University of Minnesota Press, 1957), p. 51, pp. 388-389; Antonio Delhumeau Arrecillas et al., *México: Realidad política de sus partidos* (México: Instituto Mexicano de Estudios Políticos, 1970), pp. 87-88; and Richmond, "Mexico," pp. 309-310.

125. Richmond, "Mexico," p. 311.

126. D'Antonio and Form, *Influentials in Two Border Cities,* p. 85.

127. Ugalde, *Power and Conflict,* p. 43.

128. Roderic Ai Camp, "The Cabinet and the Técnico in Mexico and the United States," *Journal of Comparative Administration* 3 (August 1971): 194.

129. Roger Charles Anderson, "The Functional Role of the Governors: Their States in the Political Development of Mexico, 1940-1964" (Ph.D. diss., University of Wisconsin, 1971), p. 303.

130. Brandenburg, *The Making of Modern Mexico*, pp. 145-146.

131. According to the Almond and Verba survey, for example, 53 percent of the Mexican respondents could not name a single leader of the PRI, while only 5 percent could name four or more party leaders. Almond and Verba, *The Civic Culture*, p. 58. The corresponding percentages for the other four countries studied were:

	Named four or more leaders (%)	Named no party leaders (%)
United States	65	16
Great Britain	42	20
Germany	69	12
Italy	36	50

132. Foster, *Tzintzuntzan*, p. 177.

133. For a discussion of authoritarian values in Mexico see Robert E. Scott, "Mexico: The Established Revolution," in *Political Culture and Political Development*, ed. Lucien W. Pye and Sidney Verba (Princeton: Princeton University Press, 1965), pp. 344-367; and Hansen, *Politics of Mexican Development*, pp. 183-193.

134. Ralph L. Beals, *Cherán: A Sierra Tarascan Village* (Washington, D.C.: Smithsonian Institution, U.S. Government Printing Office, 1946), p. 112.

135. Franz Alfred Von Sauer, "Ideological Politics in Mexico and the Partido Acción Nacional: A Case Study in Political Alienation" (Ph.D. diss., Georgetown University, 1971), pp. 161-163. Published as *The Alienated "Loyal" Opposition: Mexico's Partido Acción Nacional* (Albuquerque: University of New Mexico Press, 1974).

136. Kenneth F. Johnson, *Mexican Democracy: A Critical View* (Boston: Allyn and Bacon, 1971), pp. 133-134.

Chapter 3: Historical Aspects of the Profit-Sharing Issue

1. The specific clauses of Article 123 that were to be modified were II, III, VI, IX, XXI, XXII, and XXXI of Part A (Apartado A).

NOTES

2. Alberto Trueba Urbina and Jorge Trueba Barrera, *Ley Federal del Trabajo, reformada y adicionada,* 53d ed. (Mexico, D.F.: Editorial Porrúa, 1966), p. xxv. All translations are mine, unless otherwise indicated.

3. Ibid., p. xxvi.

4. For a detailed account of the inclusion of the profit-sharing provisions in Article 123 of the Constitution, see Octavio A. Hernández, Alfredo Uruchurtu G., Jaime Castellanos, Ernesto Valderrama Herrera, "Antecedentes legales, nacionales y extranjeros" (versión definitiva), *Memoria de la Primera Comisión* (México, D.F.: Comisión Nacional para el Reparto de Utilidades, 1964), 3:617-795; and Francisco Lerdo de Tejada, *Manual práctico de repartición de utilidades* (Buenos Aires, Bibliografía Omeba [Colección América en Letras], 1966), chap. 3.

5. Hernández, et al., "Antecedentes legales" p. 641.

6. Ibid., pp. 642-649.

7. Mario de la Cueva, *Derecho mexicano del trabajo* (México, D.F.: Editorial Porrúa, 1961), 1:141-142.

8. Adolfo López Mateos, "Iniciativa de reformas a las fracciones II, III, VI, IX, XXI, XXII y XXXI del inciso 'A' del Artículo 123 de la Constitución General de la República," *Revista Mexicana del Trabajo,* special number (April 1963), pp. 11-13. This issue contains the 1961 amendment to Article 123, the congressional debates on the amendment, the 1962 reforms to the Federal Labor Law, which were necessitated by the 1961 constitutional amendment, and the congressional debates on the labor law reforms.

9. de la Cueva, *Derecho mexicano del trabajo,* p. 692.

10. Interview with labor representative to the National Profit-Sharing Commission, 1967, Mexico City.

11. CROM, *Memoria de los trabajos realizados por el H. Comité Central durante su ejercicio del 1 de agosto de 1955 al 31 de julio de 1957,* México, D.F., p. 68.

12. Frederic Meyers, "Party, Government and the Labor Movement in Mexico: Two Case Studies" (Paper prepared for the International Institute for Labour Studies Research Conference on Industrial Relations and Economic Development, Geneva, Switzerland, 1964), mimeographed, p. 37.

13. Secretaría del Trabajo y Previsión Social, *Memoria del Congreso Mexicano de Derecho del Trabajo y Previsión Social, 19 al 23 de julio de 1949* (México, D.F., Talleres Gráficos de la Nación, 1950), 1:566-577.

14. Cámara de Diputados del XLVI Congreso de la Unión, (first year, section 7, no. 11). (Comisión Segunda de Trabajo, File 0-07-45-6/28-2-3, September 2, 1952. (Includes letter to the Chamber from Second Labor Committee dated November 30, 1964.)

15. Vicente Lombardo Toledano, "La participación de las utilidades y los intereses de la clase obrera," *Política* 4 (October 1, 1963), supp., p. xiii.

16. Ibid., p. xii.

17. *Ceteme*, July 15, 1950, pp. 4-5. (*Ceteme* is the newspaper of the CTM.)

18. *Ceteme*, February 15, 1951, p. 5.

19. Ibid.

20. The CTM reforms dealt with other maters as well. However, the profit-sharing reform was one of the most controversial.

21. *Ceteme*, December 9, 1955, p. 7.

22. Cámara de Diputados del Congreso de los Estados Unidos Mexicanos, *Diario de los debates*, Año II, Período ordinario, XLII Legislatura, vol. 1, no. 40, sessions of December 29, 1953, p. 104.

23. *Ceteme*, December 14, 1951, p. 8; and February 20, 1952, p. 1.

24. *Ceteme*, November 16, 1951, p. 1.

25. CTM Comité Nacional, *Informe al XLV Consejo Nacional Ordinario, 29, 30, 31 de julio, 1952*, México, D.F., p. 9.

26. Ibid., p. 10.

27. CROM, *Memoria de los trabajos realizados por el H. Comité Central durante su ejercicio del 1 de agosto de 1955 al 31 de julio de 1957*, México, D.F., p. 84.

28. CROM, *Memoria de los trabajos realizados por el Comité Central durante su ejercicio del 1 de agosto de 1951 al 31 de julio de 1953*, México, D.F., p. 82.

29. Rigoberto González, "Estudio sobre la participación de utilidades en relación a las fracciones VI y IX del artículo 123 constitucional," *Estudios proletarios* (México, D.F.: Ediciones CROM, 1958), 1:48, 53-54.

30. Comisión Nacional para el Reparto de Utilidades, *Memoria de la Primera Comisión*, 1:111.

31. Alfonso Alvírez Friscione, "Fundamentos axiológico-jurídicos de la participación en las utilidades y efectos sociológico-jurídicos de esta institución" (Tesis profesional, Universidad Nacional Autónoma de México, Facultad de Derecho, México, D.F., 1965), pp. 242-243.

32. See the agendas for the meetings of the National Council of the CTM, published in *Ceteme* between 1951 and 1956.

33. *Ceteme*, July 10, 1953, p. 1.

34. Cámara de Diputados, *Diario de los debates*, Año II, Período Ordinario, XLIII Legislatura, vol. 1, no. 12, session of October 23, 1956, p. 11.

35. One of the functions of a presidential campaign in Mexico is the gathering of information. During the campaign, the PRI candidate visits all parts of the country and solicits the views and opinions of individuals and groups. The Mexican Congress aids in this information-gathering process

by holding hearings on controversial proposals prior to the election of a new president. The 1952 hearings on the CTM-sponsored labor reforms that included profit sharing, for example, also coincided with a presidential campaign.

36. Congress finally dealt with the CTM proposals in 1965, fourteen years after they had first been submitted to Congress and four years after President López Mateos had sent to Congress the 1961 amendment that provided for the implementation of the profit-sharing clauses of the Constitution. The congressional committee decided in 1965 to archive the proposals since they were by then obsolete. México, Cámara de Diputados, *Diario de los debates*, Año II, Período Ordinario, XLVI Legislatura, vol. 1, no. 23, session of November 26, 1965, pp. 10-17.

37. Rigoberto González, "Estudio sobre la participación de utilidades," pp. 55-56.

38. Academia Mexicana de Derecho del Trabajo, *Memoria de la Primera Asamblea Nacional de Derecho del Trabajo, 18 al 22 de octubre de 1960* (México, D.F.), p. 255; and Academia Mexicana de Derecho del Trabajo, *Memoria de la Segunda Asamblea Nacional de Derecho del Trabajo, 15 al 22 de noviembre de 1961* (México, D.F., Talleres Gráficos de la Nación), p. 185.

39. Rigoberto González, "Estudio sobre la participación de utilidades," p. 56.

40. Academia Mexicana de Derecho del Trabajo, *Memoria de la Segunda Asamblea Nacional*, pp. 40-41.

41. *Excelsior*, June 22, 1961, p. 5.

42. *Ceteme*, November 11, 1961, pp. 1, 6.

43. This conclusion contradicts that of one analyst who stated that "the law for participation in profits seems to represent an important legal milestone in satisfying demands and meeting major issues raised by the aspirations of organized workers in the industrial sector" (L. Vincent Padgett, *The Mexican Political System* [Boston: Houghton-Mifflin Company, 1966], p. 174).

44. *El Universal*, October 13, 1951, pp. 1, 10.

45. *El Nacional*, October 13, 1951, p. 2.

46. *Ceteme*, July 31, 1953, pp. 1, 8.

47. PRI, "Dictamen sobre 'La declaracíon de principios,'" *VIII Asamblea Nacional del PRI*, 1960, March 27, 1960, p. 8 of section "Declaracion de Principios," PRI archives, Mexico City.

48. PRI, *III Asamblea Ordinaria del PRI, 1960*, Turno 10, Hoja 4, HIDALGO-MORENO, March 29, 1960, PRI archives, Mexico City.

49. Interview with CTM adviser to the labor representatives on the National Profit-Sharing Commission, 1968, Mexico City.

Another important PRI functionary expressed surprise that profit sharing was incorporated into the Declaration of Principles and not into the Program of Action. He stated that it should have been included in the Program of Action and attributed its absence to "an error" since "if the declaration and the program are not congruent, the statement of the declaration would be rendered inconsequential" (Interview with PRI functionary, 1968, Mexico City).

50. Clause 16 of the 1960 Program of Action states that the PRI would aid unions in the municipios to collect and furnish data necessary to determine the minimum wage and strive to make the minimum wage correspond to the real cost of living in each region. PRI, "Comisión Dictaminadora de Programa de Acción" *III Asamblea Nacional del PRI, 1960,* PRI archives, Mexico City.

51. PRI, "Declaración de Principios," *III Asamblea Nacional del PRI, 1960,* March 27, 1960, p. 8, PRI archives, Mexico City.

52. *La Nación,* July 2, 1961, p. 4. *La Nación* is the official organ of the PAN. This issue included a transcript of the taped television debate between the PRI and the PAN spokesmen.

53. PAN, *Plataforma que sostendrá el PAN en la campaña electoral para renovación de poderes federales en 1952 y que fué aprobada por la Convención Nacional reunida en la Ciudad de México del 17 al 20 de noviembre de 1951,* México, D.F., p. 9.

54. PAN, *Plataforma política de Acción Nacional (Aprobada en la XIII Convención del Partido — 24 de noviembre de 1957),* p. 14.

55. Interview with a PAN leader, 1968, Mexico City.

56. *Excelsior,* September 22, 1960, p. 1. The PRI's incorporation of profit sharing into its Declaration of Principles at its 1960 convention went virtually unreported, probably because it was not accompanied by proposals regarding the specific profit-sharing system desired.

57. COPARMEX, Departamento para el Estudio de la Participación de Utilidades y Salarios Mínimos, *Antecedentes nacionales,* México, D.F., 1963, pp. 81-93.

58. Interview with a PAN leader, 1968, Mexico City.

59. CROM, *Memoria de los trabajos realizados por el H. Comité Central durante su ejercicio de 1 de agosto de 1955 al 31 de julio de 1957,* México, D.F., p. 69.

60. COPARMEX, *La participación en las utilidades: Estudio de CONCAMIN, CONCANACO y COPARMEX,* no. 3, Serie Documentos y Discursos, México, D.F., July 3, 1953.

61. Euquerio Guerrero, "Participación de los trabajadores en las utilidades de las empresas," February 1961, mimeographed.

62. *Excelsior,* October 1, 1960, p. 1.

63. Juan Landerreche Obregón, *Participación de los trabajadores en las utilidades de las empresas* (México, D.F.: Editorial Jus, 1956), p. 217.

64. The following interpretation supports the conclusion of Douglas A. Chalmers regarding the diffuse nature of what he calls "agenda-setting

stimuli" in the Latin American policy-making process. Douglas A. Chalmers, "Parties and Society in Latin America" (Paper prepared for delivery at the 1968 Annual Meeting of the American Political Science Association, Washington, D.C.), mimeographed, p. 17.

65. Every other president since 1917 had first served either as secretary of the interior (gobernación) or secretary of defense. The 1958 departure from traditional recruitment patterns can be attributed to a recognition on the part of the regime's elite that it was time to placate the organized labor movement in order to quell its increasing dissatisfaction with the regime. Once president, López Mateos was expected to give visible signs of the prolabor orientation that was responsible for his elevation to the presidency.

66. *Ceteme*, September 3, 1960, p. 1.

67. Interview with a member of the commission that drafted the 1961 constitutional amendment, 1968, Mexico City. Under the administration of López Mateos's successor, Díaz Ordaz (1964-1970), draft copies of a totally reformed Federal Labor Law were circulated and opinions were solicited. Much of the work on this project had been done under the López Mateos presidency by the three-member commission. The new labor law took effect on May 1, 1970.

68. James W. Wilkie, *The Mexican Revolution: Federal Expenditure and Social Change since 1910* (Berkeley: University of California Press, 1967), p. 187; John Isbister, "Urban Employment and Wages in a Developing Economy: The Case of Mexico," *Economic Development and Cultural Change* 20 (October 1971): 24-46.

69. Pablo González Casanova, *La democracia en México* (México, Ediciones Era, 1965), p. 25. Although González Casanova does not offer an explanation for this atypical behavior, it is probably the result of persistent neglect of the organized labor movement by a series of Mexican presidents who, whether through faulty information or their own deficient judgment, underestimated the intensity of labor discontent and overestimated the ability of the co-opted labor leaders to control it. Since the combination of persistent neglect and presidential miscalculation does not occur often, rank-and-file mobilization and repudiation of leaders also does not occur frequently.

70. *Excelsior*, July 2, 1960, p. 1.

71. Wilkie, *The Mexican Revolution*, p. 32. The principal social welfare projects of the López Mateos regime included "the provision of pre-fabricated schoolhouses on a large scale, the establishment of rural health centers and the provision of 'Student Breakfasts' . . . to over half the schoolchildren in the country" (Linda S. Mirin, "Public Investment in Aguascalientes: A Study in the Politics of Economic Policy" [Ph.D. diss., Harvard University, 1964], pp. 54-55).

72. See, for example, Stanley Ross, ed., *Is the Mexican Revolution Dead?* (New York: Alfred A. Knopf, 1966).

73. For an elaboration of this idea see Murray Edelman, *The Symbolic Uses of Politics* (Urbana: University of Illinois Press, 1967), chap. 2. The Mexican regime does not officially consider the private-sector groups to be "revolutionary." They are therefore excluded from the revolutionary symbolism of the regime. Consequently, they cannot receive symbolic pay-offs. Their rewards must always be tangible. In contrast, the peasants and the workers are very much a part of the revolutionary symbolism and can therefore be rewarded both symbolically and tangibly. The proportion of symbolic to tangible rewards for groups included in the revolutionary imagery varies inversely with the level of mobilization. Thus in Mexico, the peasants, who are less mobilized than the workers, receive a greater proportion of symbolic pay-offs than do the workers.

74. *Ceteme,* July 3, 1953, p. 1.

75. According to a study of income distribution in Latin America done by the Economic Commission for Latin America in the early 1960s, income distribution in Mexico is significantly more unequal than in most other Latin American countries. The inequality results, however, "not so much [from] a high concentration in the top segments as [from] a marked disparity between the top and bottom halves of the population, although the distribution within each is more even than in [the other Latin American] countries." ("Income Distribution in Latin America," *Economic Bulletin for Latin America* 12 [October 1967]: 38-60, cited in Isbister, "Urban Employment and Wages," p. 44).

76. The most widely read critique was that of Ifigenia Navarrete, *La distribución del ingreso y el desarrollo económico de México* (México: Instituto de Investigaciones Económicas, Escuela Nacional de Economía, 1960). See also Victor L. Urquidi, "El impuesto sobre la renta en el desarrollo económico en México, *El Trimestre Económico* 23 (October-December 1956): 424-437; Benjamin Retchkiman, "Distribución del ingreso," *Revista de Economía* 21 (August 15, 1958): 224-230; Victor L. Urquidi, "La perspectiva del crecimiento económico y la repartición del ingreso nacional," *Comercio Exterior* 9 (April 1959): 198-203; Rafael Izquierdo, Leopoldo Solís, Victor L. Urquidi, "La distribución del ingreso y el desarrollo económico de México," *Comercio Exterior* 11 (February 1961): 86-90; and Victor L. Urquidi, "Problemas fundamentales de la economía mexicana," *Cuadernos Americanos* 114 (January-February 1961): 69-103.

77. See, for example, Hugo B. Margáin, *Reparto de utilidades* (México: Publicaciones Especializadas [PESA, Selección de Estudios Latinoamericanos], 1964), p. 22; the speech by Lic. Juan Sánchez Navarro,

the president of CONCAMIN in the early 1960s, delivered before a meeting of Mexican industrialists, reprinted in CONCAMIN, *Boletín Quincenal* 14 (April 1, 1963): 8; and Philippe C. Schmitter and Ernst B. Haas, *Mexico and Latin American Economic Integration* (Berkeley: University of California Press, 1964), p. 24.

78. Roger D. Hansen, *The Politics of Mexican Development* (Baltimore: Johns Hopkins Press, 1971), p. 85; Henry J. Gumpel and Hugo B. Margáin, *Taxation in Mexico* (Boston: Little, Brown and Company, 1957); and Hugo B. Margáin, "El sistema tributario," in *México: 50 años de revolución*, ed. Enrique Beltrán et al., vol. 1, *La economía*, (México: Fondo de Cultura Económica, 1960-1962), pp. 537-567.

79. Mexico has one of the lightest tax burdens among developing countries in general and within Latin America in particular. In terms of a ratio of tax revenue to GNP, Mexico's "tax effort" ranked 66 out of a total of 72 countires. Brazil, Argentina and Chile ranked 21, 27, and 23, respectively. Jorgen R. Lotz and Elliott R. Morss, "Measuring 'Tax Effort' in Developing Countries," cited in Hansen, *Politics of Mexican Development*, p. 84.

80. López Mateos did, however, make a series of minor tax reforms that increased revenues from direct taxes on corporate and personal incomes. Hansen, *Politics of Mexican Development*, p. 218.

81. Note the conclusion of one informed observer: "Politically, profit sharing would have met the same resistance as any attempt to reform the tax system. Opposition to profit sharing would have been the same as the opposition that attempts at fiscal reform always encounter. But a fiscal reform would have achieved the double objective of capturing a greater part of the national product [without necessarily modifying the tax rates] and would have been better for the development of Mexico than profit sharing" (Ignacio Pichardo, "Algunas consideraciones generales sobre la participación de utilidades," *Comercio Exterior* 14 [January 1964]: 11).

82. Interview with the president of the National Profit-Sharing Commission, 1968, Washington, D.C.

83. Ibid.

84. Ibid.

85. The government's capabilities with respect to income-tax collection have been continually increasing. When the income tax was created in 1924, the money it produced equaled 1 percent of the total amount collected by the government. In 1957, it equalled 33.76 percent. Margáin, "El sistema tributario," p. 559. Thus, over the years the Mexican government has placed decreasing reliance on indirect taxes (i.e., sales and excise taxes, export taxes), which require fewer control capabilities, and increasing reliance on the more difficult to collect direct taxes (i.e., income tax). In fact, the percentage of government revenue resulting from direct

taxation is among the highest of all Latin American countries. In a 1960 study, only Chile (32.8 percent direct taxes) ranked higher than Mexico's 31.5 percent. Most of the other countries derived between 10 and 17 percent of their revenues from direct taxation. Raynard M. Sommerfeld, *Tax Reform and the Alliance for Progress* (Austin: University of Texas Press, 1966), p. 49.

86. Agustín Reyes Ponce, "Estudio sobre la participación legal de los trabajadores en las utilidades de las empresas," June 9, 1962, mimeographed, p. 4.

87. Interview with a private-sector representative to the National Profit-Sharing Commission, 1968, Mexico City.

88. *Ceteme*, August 29, 1952, p. 3.

89. *Ceteme*, September 26, 1952, p. 3.

90. COPARMEX, Boletín #SPCP-60/64, Departamento de Relaciones Públicas, Servicio de Prensa, May 2, 1964.

91. The disruptive strikes of 1957-1959 do not contradict this conclusion, for they were directed against the government, not against the private sector. The petroleum industry, the railroads, and the telegraph system are all government-owned.

Chapter 4: Government Initiation and Interest-Group Response

1. Interview, 1968, Mexico City. The status of the informant cannot be revealed.

2. Deliberations do not always reinforce the Mexican president's commitment to a decision. They also can result in the rejection of the decision or in an inconclusive division of opinion. In the latter case, the president will have a trusted subordinate publicly propose the idea for the decision. If the proposal is favorably received by the principal groups, the president will publicly announce the decision. If the proposal generates substantial opposition, the president will only publicly associate himself with it if he is both willing and able to devote substantial time and resources to the demobilization of the proposed decision's opponents. The president himself does not publicly "test" decisions to which he is committed because public association of the president with a proposed decision is tantamount to making the decision.

Two examples of proposed decisions that were privately deliberated, publicly tested, and then rejected were the idea to democratize (i.e., decentralize) the PRI and the idea to allow for the consecutive reelection of members of the Chamber of Deputies. The democratization idea was tested by the PRI's president, Carlos Madrazo, in 1965. The consecutive reelection idea was tested by the PRI's leader in the Chamber of Deputies,

Alfonso Martínez Domínguez, between 1964 and 1965. Both "tests" generated substantial dissension within the ruling coalition, with the result that the Mexican president refrained from implementing either decision.

3. The members of the commission were Dr. Mario de la Cueva, one of Mexico's foremost authorities on labor legislation, Lic. Ramiro Lozano, a labor lawyer who had worked with and headed a federal junta of conciliation and arbitration for many years, and Lic. María Salmorán de Tamayo, another labor lawyer who had worked with and had headed a junta of conciliation and arbitration. Interviews with members of the commission, 1968, Mexico City.

4. Ibid.

5. A prominent businessman, during my 1968 interview with him in Mexico City, suggested a provocative corollary. He believes that if the government publicizes its intention to undertake a controversial reform, it may be an indication that the reform will not occur.

6. *Excelsior*, June 22, 1961, p. 5.

7. Adolfo López Mateos, *Pensamiento y programa* (México, D.F.: Editorial La Justicia, 1961).

8. Adolfo López Mateos, *Pensamiento en acción* (México, D.F.: Ediciones de la Oficina de Prensa de la Presidencia de la República, 1964), 2:10.

9. *Excelsior*, December 29, 1961, p. 9.

10. *El Nacional*, section 2, December 31, 1961, p. 4.

11. *Excelsior*, January 2, 1961, p. 13. The fact that the reforms surprised the private sector was confirmed by Lic. Guajardo Suárez, the head of COPARMEX, during my interview with him in 1968 in Mexico City.

12. Interview with a labor representative on the National Profit-Sharing Commission, 1968, Mexico City.

13. Interview with an adviser to the labor representatives on the National Profit-Sharing Commission, 1968, Mexico City.

14. *Ceteme*, November 11, 1961, pp. 1, 6.

15. Estados Unidos Mexicanos, Comisión Nacional para el Reparto de Utilidades, *Memoria de la Primera Comisión*, vol. 1, México, 1964, p. 679.

16. *Revista Mexicana del Trabajo*, Special number, 10 (April 1963): 15.

17. Ibid.

18. Estados Unidos Mexicanos, *Memoria de la Primera Comisión:* I: 679.

19. The principal addition that several labor deputies wanted to make was a statement that the right of the workers to share in profits "in no way limits the other rights that the workers have won" (Estados Unidos Mexicanos, *Memoria de la Primera Comisión*, 1:703).

20. *El Nacional,* March 31, 1961, p. 3.

21. *El Nacional,* May 26, 1961, p. 1.

22. Albert O. Hirschman, *Journeys Toward Progress: Studies of Economic Policy-Making in Latin America* (New York: Twentieth Century Fund, 1963), p. 230.

23. Raymond Vernon, *The Dilemma of Mexico's Development* (Cambridge, Mass.: Harvard University Press, 1963), p. 122.

24. Note, for example, the editorial in *El Nacional* headlined "The Businessmen Have Understood Their Duty." This editorial congratulates the private sector for its "about face in favor of social justice" (*El Nacional,* November 10, 1961, p. 3).

25. The headline of the article in *El Nacional* reporting the president's New Year's message included the phrase "a special call to the private sector" (*El Nacional,* January 1, 1962, p. 1). Three weeks later, *El Nacional* reported that, for the first time, a Mexican president had addressed the annual assembly of the CNIT. *El Nacional,* January 25, 1961, pp. 1, 4.

26. CONCAMIN, Circular #62/95, July 6, 1962.

27. *Ceteme,* June 16, 1962, p. 1, reported the invitation to the CTM. The government's invitation to the CROM was reported in CROM, *Memoria de la CROM, 1961-1963* (México, D.F., 1963), p. 124. The FOR indicated that its opinions would be given to the government in *Engrane,* November 15, 1962, p. 1.

28. Interview with a labor representative on the National Profit-Sharing Commission, 1968, Mexico City.

29. Although other groups (e.g., societies of accountants, lawyers or economists) might have appreciated an opportunity to express their views on profit sharing to the secretary of labor, the government only invited the most relevant parties (i.e., the labor and business organizations) to do so. The concept of relevant parties is related to the corporate principles embodied in the Mexican Constitution. The Mexican elite does not consider itself obligated to give equal consideration to the viewpoints of all citizens with regard to a particular issue. Instead, only those groups to be affected by a decision will be specifically contacted should the president desire such contact. See, for example, the article by a high-ranking bureaucrat in the Ministry of Labor, Francisco Rostro P., "Stability, Development and the Private Sector," *El Día,* August 20, 1967, p. 5.

The functional groups apparently share the governing elite's conception and do not expect to be consulted except on matters of specific concern to them. As an official of the Bankers' Association stated when I interviewed him, "before the government sends legislation to Congress, it is shown to the *involved groups* and they are allowed to make comments. Sometimes their comments are taken into consideration and sometimes they are not. Usually, when they are not, the government is right and is doing what it has to do" (Interview, 1967, Mexico City). Italics mine.

30. *El Nacional,* October 3, 1962, p. 1.

31. Letter from a representative of the private sector addressed to CONCAMIN President Lic. Juan Sánchez Navarro, dated August 31, 1962, in folder entitled "Reformas al Artículo 123," CONCAMIN headquarters, Mexico City; and CONCANACO, *XLV Asamblea General Ordinaria,* September 1962, p. 27.

32. Letter from a representative of the private sector dated August 31, 1962, in "Reformas al Artículo 123," CONCAMIN headquarters, Mexico City.

34. They were: new firms, during the first two years of existence; new firms producing a new product, during the first four years of existence; firms involved in extractive industry, during the period of exploration; private charitable institutions; the Mexican Institute of Social Security and public decentralized institutions with charitable, welfare, or cultural ends; firms that possess less capital than the minimum set by the Ministry of Labor and Social Welfare. Alberto Trueba Urbina and Jorge Trueba Barrera, *Ley Federal del Trabajo, reformada y adicionada,* 53rd ed. (México: Editorial Porrúa, 1966), p. 46.

35. See the *Revista Mexicana del Trabajo,* special number (April 1963) for the original 1962 reforms to the Federal Labor Law, before they were partially amended by the Chamber of Deputies. For the amended Labor Law reforms see Trueba Urbina and Trueba Barrera, *Ley Federal del Trabajo.*

36. The opinions which the private sector representatives presented to the Ministry of Labor can be found in Estados Unidos Mexicanos, Comision Nacional para el Reparto de Utilidades, *Memoria de la Primera Comisión,* "Estudio sobre las bases para reglamentar la fracción IX del artículo 123 constitucional," June 1962 (México, D.F., 1964), 3:827-837.

The views of the labor sector cited here are essentially those of the CTM. At an extraordinary assembly of the BUO, the delegates voted to make the CTM views their own and present them to the secretary of labor. I was unable to learn the specific opinions the CNT, the rival labor confederation, expressed to the government. For the CTM and BUO views, see *Ceteme,* November 10, 1962, p. 3; and *Ceteme,* November 17, 1962, p. 3.

37. The report of the congressional committees that studied the 1962 reforms and the congressional debates on the 1962 labor law reforms can be found in the *Revista Mexicana del Trabajo,* special number (April 1963), pp. 79-146.

38. When President López Mateos had submitted his amendment to Article 123 of the Constitution to Congress in December 1961, neither the

Senate nor the Chamber of Deputies had changed the amendment in any way. However, Congress, particularly the Chamber of Deputies, obviously considered itself to be under fewer constraints in the case of the 1962 reforms. A constitutional amendment involves the Constitution of 1917, the basis of the legitimacy of the Mexican regime and an important element in the regime's consensus. The Federal Labor Law, on the other hand, is imbued with none of the symbolic importance of the Constitution.

39. Note a bulletin released by COPARMEX's Department of Public Relations soon after the constitutional amendment became public knowledge. It stated, "Since profit sharing is a right granted since 1917 by Mexican law, the general bases that [the amendment] proposes only advance and facilitate the enforcement of that right" (COPARMEX, Departamento de Relaciones Públicas, Servicio de Prensa Confederación Patronal, Boletín #SPCP-31/62, January 1962).

40. CONCAMIN, Draft of Circular 62, January 1962 (never sent), in "Reformas al Artículo 123," CONCAMIN headquarters, Mexico City. Note that CONCAMIN was under no illusions regarding the government's influence over the leaders of the labor movement.

41. *Excelsior*, January 2, 1962, p. 13.

42. *Siempre!* 446 (January 10, 1962): 9.

43. CONCAMIN, Draft of Circular 62, January 1962 (never sent), in "Reformas al Artículo 123."

44. *Excelsior*, December 29, 1961, pp. 1, 9.

45. CONCAMIN, Circular #62/95, July 6, 1962.

46. Letter from a high-ranking CONCAMIN official to heads of CONCANACO, the Bankers' Association, and COPARMEX, dated January 16, 1962, in "Reformas al Artículo 123."

47. *Revista Bancaria* 5 (July-August 1962): 357.

48. Sources for reconstruction of the membership of the private sector's gran comisión include letters from a high-ranking CONCAMIN official addressed to various members of CONCANACO, COPARMEX, and the Bankers' Association, dated February 13, 1962 to May 31, 1962, in "Reformas al Artículo 123."

49. CONCAMIN, Circular #62/26, February 2, 1962.

50. Interview with a private-sector representative to the National Profit-Sharing Commission, 1968, Mexico City.

51. CONCAMIN, Circular 62/95, July 6, 1962.

52. Letter from CONCAMIN President Lic. Juan Sánchez Navarro to Lic. Agustín Reyes Ponce, dated March 12, 1962, in "Reformas al Artículo 123."

53. CONCAMIN, Circular 62/95, July 6, 1962.

54. Letter from CONCAMIN President Sánchez Navarro to the

secretary of labor asking for extension, dated June 6, 1962; letter from the secretary of labor to Sánchez Navarro, extending deadline, dated June 11, 1962, in "Reformas al Artículo 123."

55. COPARMEX, Departmento de Relaciones Públicas, Circular #178, June 30, 1962.

56. Minutes of the June 12, 1962 meeting of the gran comisión, in "Reformas al Artículo 123."

57. Ibid.

58. Ibid.

59. CONCAMIN, Circular #62/95, July 2, 1962.

60. "Estudio que presenta el sector de empresarios al Sr. Lic. Salomón González Blanco, Secretario del Trabajo y Previsión Social, respecto de las bases para reglamentar la fracción IX del Artículo 123 Constitucional," mimeographed (circa June 1962).

61. CONCAMIN, Circular #62/106, July 18, 1962.

62. COPARMEX, Departamento de Relaciones Públicas, Circular #197, August 21, 1962 and Circular #198, August 21, 1962.

63. COPARMEX, Departamento de Relaciones Públicas, Servicio de Prensa Confederación Patronal, Boletín SPCP-43/62, October 1962.

64. COPARMEX, *Voz Patronal* 10 (October 1962-January 1963): 1-3.

65. This statement came from a leader of the FOR, the newest and most left-wing labor group. The FOR had never in any way indicated a desire for profit sharing. *El Universal*, December 29, 1961, p. 10.

66. This quote refers to a series of three conferences sponsored by the FROC, the Mexico City branch of the CROC, held in February 1962. *El Nacional*, February 12, 1962, p. 1.

67. On December 23, 1962, for example, *Excelsior* quoted the head of the Electricians' Union as saying that "profit-sharing will not weaken [*aplacar*] the class struggle." During a round table discussion sponsored by the FOR in November 1962 the head of the CNT "rejected the employers' thesis that profit sharing converts workers into partners of their employers" (*El Día*, November 29, 1962, p. 1).

68. Some of the private-sector leaders, in order to combat accusations that they were against the establishment of a profit-sharing system, made public statements to the effect that they favored profit sharing because it would facilitate collaboration between workers and employers. The head of COPARMEX, for example, was quoted as stating that "profit sharing will favor the harmony and collaboration between capital and labor" and "reduce and help to prevent the class struggle" (*El Universal*, November 24, 1962, p. 1). His statement obligated Fidel Velázquez, the head of the very moderate CTM, to "strongly dispute" the statement. Velázquez claimed that not only would the class struggle not end with profit sharing, but that the workers would reaffirm the class struggle. *Excelsior*, November 25, 1962, p. 1.

69. The employers' center of the Federal District had withdrawn from COPARMEX in 1951 because it felt that COPARMEX's moderation was no longer representative of its own more conservative views.

70. *Engrane*, November 1, 1962, p. 7. *Engrane* is the newspaper of the FOR.

71. *Ceteme*, November 17, 1962, p. 1; *Engrane*, November 15, 1962, p. 1.

72. *Ceteme*, November 3, 1962, pp. 1, 3, reported the October 31, 1962 meeting between the BUO leaders and the secretary of labor. *El Nacional* (October 31, 1962, p. 8) reported the similar meeting between the CNT head and the secretary of labor.

73. An *Excelsior* article (January 21, 1962, pp. 4-5), for example, reported how the PRI responded to COPARMEX's attacks and "defended" the reforms to Article 123. An article in *El Nacional* (January 20, 1962, pp. 1-2) spoke of the PRI's defense of the profit-sharing reforms in the face of attacks on them by the Employers' Center of the Federal District.

74. The activities of the minor parties during 1962 were limited to the expression of general praise for both the constitutional amendment and the Federal Labor Law reforms by their few representatives in the Chamber of Deputies. The representatives of the minor parties voted in favor of both, as did the PRI deputies.

75. Robert E. Scott, *Mexican Government in Transition* (Urbana: University of Illinois Press, 1959), p. 18.

Chapter 5: The Politics of the National Profit-Sharing Commission

1. See the Appendix for a list of the members of the National Profit-Sharing Commission.

2. Interview with the president of the National Profit-Sharing Commission, 1968, Washington, D.C.

3. Interviews with members of the technical staff of the National Profit-Sharing Commission, 1968, Mexico City.

4. Raymond Vernon, *The Dilemma of Mexico's Development* (Cambridge, Mass.: Harvard University Press, 1963), pp. 137-138. Miguel S. Wionczek, in an essay on the evolution of economic planning in Mexico, also mentions the delegation of considerable decision-making power by the politicians to the technocrats on technical, economic matters. See his "Incomplete Formal Planning: Mexico," in *Planning and Economic Development*, ed. Everett Hagen (Homewood, Ill.: Irwin Press, 1963), p. 178.

5. The five branches of industry were:
   1. Mining, hydrocarbons, petrochemicals, metallurgy.
   2. Transportation, telephones, television and radio, electrical

firms, credit and insurance institutions, hotels, restaurants, educational institutions, hospitals, barbershops, beauty parlors, sanitation, food preparation.
3. Manufacturing industries.
4. Commercial firms.
5. Nonmanufacturing industries and activities (i.e., construction, agriculture, fishing).
6. The first method was used by COPARMEX (COPARMEX, Circular Extraordinaria #1), some time between January 21 and February 8, 1963). The second method was employed by CONCANACO (CON-CANACO, Circular #46-46, January 30, 1963). The third method was used by CONCAMIN (CONCAMIN, Circular 63/23, January 30, 1963).
7. The executive committee of the Bankers' Association decided not to elect representatives of the banking sector to the National Profit-Sharing Commission. *Revista Bancaria* 3 (April 1963): 207. Several members of the private sector suggested to me that this decision was taken because the Bankers' Association has a poor public image. As a result, it is usually eager to avoid publicity and participation in controversial issues that might give its opponents ammunition.
8. Members of the Mexican private sector are not, of course, involved exclusively or forever with either industrial or commercial enterprises. Heriberto Vidales, for example, was the head of a large chain of supermarkets in 1963, when he also served as the president of CONCANACO. A short time later, he became the director of Goodrich-Euzkadi, which was affiliated with CONCAMIN, and he then served as one of CONCAMIN's vice-presidents on several other occasions.
9. Interview with a representative of the private sector on the National Profit-Sharing Commission, 1968, Mexico City.
10. These firms were Fundidora de Hierro y Acero de Monterrey (Campillo Sainz), Química General (Isoard), I.E.M. (Alatorre), Goodrich-Euzkadi (Vidales), and CONDUMEX (García Sainz).
11. *Ceteme*, February 2, 1963, p. 1.
12. Interview with a representative of organized labor to the National Profit-Sharing Commission, 1968, Mexico City.
13. *Ceteme*, March 2, 1963, p. 4.
14. CTM, *Informe del Comité Nacional: LXVII Asamblea General Ordinaria del Consejo Nacional, agosto 30, 31, septiembre 1, 2, de 1963*, México, D.F., pp. 9-10.
15. *Engrane*, March 1, 1963, p. 1.
16. Interviews with representatives of organized labor to the National Profit-Sharing Commission, 1968, Mexico City.
17. *El Nacional*, February 20, 1963, pp. 1, 7.
18. Interviews with members of the technical staff of the National Profit-Sharing Commission, 1968, Mexico City.

19. Interviews with members of the technical staff of the National Profit-Sharing Commission, 1968, Mexico City.

20. Interviews with members of the technical staff of the National Profit-Sharing Commission, 1968, Mexico City. Members of the technical staff frequently made statements like these: "Mexico, at the time of the profit-sharing decision, was more concerned with economic growth." "Economic factors were more important than political factors in drawing up the profit-sharing system." "It was very important not to damage the economic development of the country." "The members of the technical staff had very little preoccupation with political concerns; they were concerned with economic, fiscal, and legal matters."

21. Vernon, *The Dilemma of Mexico's Development*, p. 88.

22. Ibid. The succession of presidents began with Ávila Camacho, who was followed by Alemán, Ruíz Cortines, and López Mateos. For a similar conclusion, see James W. Wilkie, *The Mexican Revolution: Federal Expenditure and Social Change since 1910* (Berkeley: University of California Press, 1967), p. 277.

23. Raúl Prebisch, "El desarrollo económico de la América Latina y algunos de sus principales problemas," *El Trimestre Económico* 16 (July-September 1949): 347-431. For a summary of economic thought in Latin America from 1930 to 1950 see Felipe Pazos, "Veinte años de pensamiento económico en la América Latina," *El Trimestre Económico* 20 (October-December 1953): 552-571.

24. Vernon, *The Dilemma of Mexico's Development*, pp. 142-144.

25. Hugo B. Margáin, *Reparto de utilidades* (México, D.F.: Publicaciones Especializadas, Selección de Estudios Latinoamericanos, 1964), p. 27.

26. Interview with the president of the National Profit-Sharing Commission, 1968, Washington, D.C.

27. Interview with the president of the National Profit-Sharing Commission, 1968, Washington, D.C., and interview with an adviser to the private sector's representatives to the National Profit-Sharing Commission, 1968, Mexico City.

28. Government-created tripartite commissions are a common phenomenon in Mexico, especially when their work involves the private sector and organized labor, probably for the reasons just cited. For example, such commissions have been utilized for minimum-wage decisions and collective bargaining. The council of the Mexican Institute of Social Security is also tripartite.

29. Estados Unidos Mexicanos, Comisión Nacional para el Reparto de Utilidades, *Memoria de la Primera Comisión* (México, D.F., 1964), 1:10-11.

30. Among the groups that ultimately submitted studies to the commissions were the Mexican Association of Insurance Institutions, the

College of Public Accountants of Guadalajara, the Platform of Mexican Professions, and the National Banking Commission. For a complete list of the studies presented to the commission and the groups that sponsored them, see Estados Unidos Mexicanos, *Memoria de la Primera Comisión*, vol. 3 (general index).

31. *El Nacional*, July 26, 1963, p. 1.

32. Estados Unidos Mexicanos, *Memoria de la Primera Comisión*, 1:166-174.

33. Ibid., pp. 142-143.

34. Ibid., pp. 867-869.

35. Ibid., p. 336.

36. Interview with the president of the National Profit-Sharing Commission, 1968, Washington, D.C.

37. Interview with the president of the National Profit-Sharing Commission, 1968, Washington, D.C.

38. Estados Unidos Mexicanos, *Memoria de la Primera Comisión*, 1:271-275.

39. Ibid., pp. 456, 514.

40. Interviews with labor representatives to the National Profit-Sharing Commission, 1968, Mexico City.

41. Estados Unidos Mexicanos, *Memoria de la Primera Comisión*, 1:329-330.

42. Hugo B. Margáin, *Reparto de utilidades*, p. 32. Another reason Latin American profit-sharing systems had failed, according to Margáin, was that they allowed a deduction from profits of a certain percentage of the total capital investment of a firm, which was essentially the system demanded by the Mexican private sector's representatives.

43. Interview with the president of the National Profit-Sharing Commission, 1968, Washington, D.C.

44. These events were reconstructed on the basis of information I obtained during interviews with individuals involved with the National Profit-Sharing Commission.

45. Interviews with members of the National Profit-Sharing Commission, 1968, Mexico City and Washington, D.C.

46. Interviews with representatives of the private sector of the National Profit-Sharing Commission and members of the government's technical staff, 1968, Mexico City.

47. Estados Unidos Mexicanos, *Memoria de la Primera Comisión*, 1:520-527.

48. Ibid., pp. 520-527.

49. Ibid., pp. 528-570. The draft of the resolution outlining the profit-sharing system appears on pages 528-549.

50. Interview with the president of the National Profit-Sharing Commission, 1968, Washington, D.C.

51. Margáin's skill as head of the comission subsequently received formal recognition when President Díaz Ordaz named him ambassador to the United States. Three months before the end of his term, Díaz Ordaz named Margáin minister of the Treasury. He retained that position when President Luis Echeverría Álvarez took office. In June 1973, Margáin suddenly resigned for "reasons of health." There are two contradictory interpretations of his resignation. One attributes it to his efforts to reform and modernize Mexico's tax system, efforts that were vetoed by President Echeverría. The other explanation suggests that Margáin, because of his close links with the private sector, "may have been less than enthusiastic about some of Echeverría's plans for breathing life back into the Mexican Revolution" (*Latin America* [London] 7 [June 8, 1973]: p. 1).

52. As one of the private sector's representatives to the National Profit-Sharing Commission stated, "[The government's representatives] were interested in keeping the amount of profits distributed low for the benefit of the economy. To get a unanimous vote they perhaps had to make the amount a bit lower than they intended. But the técnicos were in no way planning to give the workers very much, so business and the técnicos were not as far apart as many people think" (Interview, 1968, Mexico City).

53. Mario de la Cueva, "El sistema mexicano para la participación de los trabajadores en las utilidades de las empresas," *El Día*, May 4, 1968, p. 10.

54. COPARMEX, *Noticia histórica sobre el Departamento para el Estudio de la Participación de Utilidades y los Salarios Mínimos*, México, D.F., pp. 3-5.

55. Cuestionario para las Confederaciones, las Cámaras de Comercio e Industria y Organizaciones de Empresarios, formulado por los representantes empresariales ante la Comisión para el Reparto de Utilidades," mimeographed.

56. COPARMEX, *Noticia histórica sobre el Departamento para el Estudio de la Participación de Utilidades y los Salarios Mínimos*, p. 6.

57. The main reasons for their failure, according to the COPARMEX study, were "the inadequate economic and social conditions" of these countries, including high levels of illiteracy, low levels of cooperation between workers and employers, and badly functioning economies. COPARMEX, Departamento para el Estudio de la Participación de Utilidades y Salarios Mínimos, *Antecedentes legales, nacionales y extranjeros. Participación obligatoria de utilidades a los obreros en los países latinoamericanos*," October 1963, México, D.F., pp. 1-3.

58. COPARMEX, Departamento para el Estudio de la Participación de Utilidades y Salarios Mínimos, *Estudio sobre los sistemas de aplicación de utilidades en favor de los trabajadores,* September 1963, México, D.F., p. 10.

59. See the index at the end of volume 3 of the Estados Unidos Mexicanos, *Memoria de la Primera Comisión* for a list of the studies presented to the National Profit-Sharing Commission.

60. *El Nacional,* December 10, 1963, p. 1.

61. Interviews with representatives of the private sector on the National Profit-Sharing Commission, 1968, Mexico City.

62. Rafael Izquierdo, "Protectionism in Mexico" in *Public Policy and Private Enterprise in Mexico,* ed. Raymond Vernon (Cambridge, Mass.: Harvard University Press, 1964), pp. 259, 277.

63. The term "redistributive" is used in the sense Theodore Lowi uses it. Lowi divided issues into three categories, redistributive, distributive, and regulatory, and hypothesized that "issues that involve redistribution cut closer than any others along class lines and activate interests in what are roughly class terms." He further stated that "if there is ever any cohesion *within* the peak associations it occurs on redistributive issues" (Theodore J. Lowi, "American Business, Public Policy, Case Studies and Political Theory," *World Politics* 16 [July 1964]: 707). Italics mine.

Lowi does not, however, deal with the question of cohesion or its absence *among* peak organizations representing interests of the same class. With regard to profit sharing, for example, which Lowi would classify as a redistributive issue, not only was there relative cohesion *within* CONCAMIN, CONCANACO, and COPARMEX, but there was also a great deal of cohesion *among* CONCAMIN, CONCANACO, and COPARMEX.

64. The term "multisector" is that of Constantine Menges. Menges pointed out that redistributive policies might affect several economic sectors (multisector, e.g., general wage increases), or only one (single-sector, e.g., agrarian reform). Constantine C. Menges, "Public Policy and Organized Business in Chile: A Preliminary Analysis," *Journal of International Affairs* 2 (1966): 358-359.

65. In his article, Menges hypothesized that redistributive policies that are multisector produce "perhaps active [but] definitely passive cohesion" among the peak associations that represent interests of the same class. Ibid., pp. 358-359. He did not distinguish between the perceived beneficiaries and those who suffer as a result of such policies, and therefore, was unable to specify whether active or passive cohesion would occur.

66. See *Excelsior,* December 11, 1963, p. 20, for labor's advertisement.

67. Estados Unidos Mexicanos, Cámara de Diputados, *Los presidentes de México ante la nación: Informes, manifiestos y documentos de 1921 a 1966*, vol. 4 (México D.F.: Editado por XLVI Legislatura de la Cámara de Diputados, 1966), p. 826.

68. *El Nacional*, December 20, 1964, p. 4.

69. *Excelsior*, December 14, 1963, p. 18.

70. *El Día*, December 14, 1963, p. 3.

71. Ibid., p. 3.

72. Interviews with labor representatives on the National Profit-Sharing Commission, 1968, Mexico City.

73. Frederic Meyers made the first estimate cited here in "Party, Government and the Labor Movement in Mexico: Two Case Studies" (Paper prepared for the International Institute for Labour Studies Research Conference on Industrial Relations and Economic Development, Geneva, August 23-September 4, 1964), mimeographed, p. 46. The higher estimate is from Ignacio Pichardo, "Algunas consideraciones generales sobre la participación de utilidades," *Comercio Exterior* (January 1964): 12.

74. *Ceteme*, June 6, 1964, p. 1.

75. *Ceteme*, June 13, 1964, p. 1.

76. *Excelsior*, February 28, 1965, p. 1.

77. Interview with a functionary of the Department of Profit-Sharing and Minimum Wages of the Ministry of Labor, 1967, Mexico City.

78. Estados Unidos Mexicanos, *Diario de los debates de la Cámara de Diputados del Congreso de los Estados Unidos Mexicanos*, Período Ordinario, XLVI Legislatura, Año II, vol. 1, no. 6, session of September 21, 1965, p. 8.

79. CTM, "Convocatoria a la Reunión Nacional del Trabajo y Seguridad Social," May 11-12, 1965.

80. An enterprise falls under federal jurisdiction: (1) when it operates in federal zones or territorial waters; (2) when the government directly or indirectly administers the enterprise; (3) when the enterprise is under a government contract or is operating as a result of a federal concession (e.g., road construction firms); (4) when the activities of a firm are regulated by a "contrato ley del trabajo," a general labor contract for the entire industry (the textile, rubber, sugar and alcohol industries are regulated by a contrato ley del trabajo). Interview, functionary of the Department of Associations of the Ministry of Labor, 1967, Mexico City.

81. Estados Unidos Mexicanos, Secretaría del Trabajo y Previsión Social, *Memoria de Labores, enero a diciembre de 1964* (México, D.F.: Talleres Gráficos de la Nación, 1965), pp. 81-83; and interview with the head of the Department of Profit-Sharing and Minimum Wages of the Ministry of Labor, 1967, Mexico City.

NOTES

82. Estados Unidos Mexicanos, Secretaría del Trabajo y Previsión Social, *Memoria de Labores, septiembre de 1970/agosto de 1971* (México, D.F.: Talleres Gráficos de la Nación, 1971), pp. 19-20.

83. Interview with the head of the Profit-Sharing Department of the Treasury, 1967, Mexico City.

84. "El fisco y la participación de utilidades," *Investigación Fiscal* 20 (August 1967): 7-8.

85. Article 125, Clause I of Chapter VIII of the *Anteproyecto, Ley Federal del Trabajo*, p. 45.

86. Ibid., Article 582, Clause V of Chapter IX, p. 232.

87. Ibid., Article 446, Clause V of Chapter II, p. 141.

88. *El Día*, February 8, 1969, p. 1.

89. *Ceteme*, March 1, 1969, p. 2.

90. The specific wording was: "Empleados de confianza (other than directors, administrators, and general managers) will participate in the profit-sharing system, but if the salary they receive is greater than that which corresponds to the highest paid worker in the plant, the latter salary, augmented by 20 percent, will be considered the maximum salary of the former" (Alberto Trueba Urbina and Jorge Trueba Barrera, *Nueva Ley Federal del Trabajo Reformada* [México: Editorial Porrúa, 1972], p. 72).

91. According to the Federal Labor Law (as amended in 1962), which was in effect until May 1, 1970, a Second National Profit-Sharing Commission could not be convened until ten years after the resolution of the First National Profit-Sharing Commission had been announced. The First National Profit-Sharing Commission Resolution was announced on December 13, 1963. In 1971, Fidel Velázquez of the CTM and other labor leaders began calling for an early formation of the Second National Profit-Sharing Commission. *Ceteme*, June 12, 1971, p. 1.

92. The first head of the Profit-Sharing Department of the Treasury estimated that in 1964, 700 million pesos were distributed among the workers. In 1965, this amount increased to 850 million and in 1966 it reached approximately 1 billion pesos. Interview, 1967, Mexico City. In 1972, when I asked the head of the Treasury's Profit-Sharing Department for the 1967-1971 figures, he claimed that these data were not available. Interview, 1972, Mexico City.

The difficulty of obtaining figures for recent years may be related to the impending constitution of the Second National Profit-Sharing Commission. The government may be withholding data in order to maximize its decision-making leeway on the commission and to avoid unnecessary confrontations. This pattern of behavior apparently exists in the educational planning process. See the study by Guy Benveniste, *Bureaucracy and National Planning: A Sociological Case Study in Mexico* (New York: Praeger Publishers, 1970), p. 96.

93. Interview with a COPARMEX economist, 1972, Mexico City. If the increase in the cost of living between 1966 and 1970 (approximately 42 pesos per year) is taken into account, the amount of money received by an average worker increased by approximately U.S. $20 between 1966 and 1970.

94. See, for example, the speech by Fidel Velázquez printed in *Ceteme*, June 12, 1971, p. 1.

95. Article 586 states: "The resolution [of the commission] will set the percentage that the workers are entitled to based on the taxable income, without making any deduction or differentiating among types of businesses" (Trueba Urbina and Trueba Barrera, *Nueva Ley Federal del Trabajo Reformada*, p. 292).

Chapter 6: The Dynamics of the Authoritarian Decision-Making Process

1. Philippe C. Schmitter and Ernst B. Haas, *Mexico and Latin American Economic Integration* (Berkeley: Institute of International Studies, University of California, 1964); Terry L. McCoy, "A Paradigmatic Analysis of Mexican Population Policy," in *The Dynamics of Population Policy in Latin America*, ed. Terry L. McCoy (Cambridge, Mass.: Ballinger Publishing Co., 1974), pp. 377-408; Guy Benveniste, *Bureaucracy and National Planning: A Sociological Case Study in Mexico* (New York: Praeger Publishers, 1970); and Martin Harry Greenberg, *Bureaucracy and Development: A Mexican Case Study* (Lexington: D.C. Heath and Company, 1970).

2. This section borrows extensively from Susan Kaufman Purcell, "Decision-Making in an Authoritarian Regime: Theoretical Implications from a Mexican Case Study," *World Politics* 26 (October 1973): 46-51.

3. Benveniste, *Bureaucracy and National Planning*, p. 13. Examples will be footnoted only when direct quotations are involved. All subsequent references to the LAFTA decision, the family planning decision, the educational planning process, and the functioning of the Ministry of Hydraulic Resources are from Schmitter and Haas, *Mexico and Latin American Economic Integration*; McCoy, "Mexico"; Benveniste, *Bureaucracy and National Planning*; and Greenberg, *Bureaucracy and Development*. References to the Spanish policy-making process are exclusively from Charles Anderson, *The Political Economy of Modern Spain: Policy-Making in an Authoritarian System* (Madison: University of Wisconsin Press, 1970).

4. Greenberg, *Bureaucracy and Development*, p. 127.

5. Anderson, *Political Economy of Modern Spain*, p. 244.

6. McCoy, "Mexico," pp. 391-329.

7. Anderson, *Political Economy of Modern Spain*, pp. 69, 243.

8. Ibid., pp. 243-244.

9. Schmitter and Haas, *Mexico and Latin American Economic Integration*, p. 24.

10. Anderson, *Political Economy of Modern Spain*, p. 5.

11. Peter Bachrach and Morton S. Baratz, *Power and Poverty: Theory and Practice* (New York: Oxford University Press, 1970), p. 43.

12. Anderson, *Political Economy of Modern Spain*, p. 240.

13. Ibid., p. 62.

14. It might be hypothesized that as the technical capabilities of the Mexican regime have increased, it has become more feasible to isolate technical from political issues. The technical problems can be resolved with less guesswork and more accurate data, which reduces the influence of politics on such issues.

15. David Ronfeldt, *Atencingo: The Politics of Agrarian Struggle in a Mexican Ejido* (Stanford: Stanford University Press, 1973).

16. Frank Brandenburg, *The Making of Modern Mexico* (Englewood Cliffs, N.J.: Prentice Hall, Inc., 1964), p. 291.

Bibliography

Public Documents

Academia Mexicana de Derecho del Trabajo. *Memoria de la Primera Asamblea Nacional de Derecho del Trabajo, 18 al 22 de octubre de 1960.* México, D.F., 1961.
———. *Memoria de la Segunda Asamblea Nacional de Derecho del Trabajo, 15 al 22 de noviembre de 1961.* México, D.F.: Talleres Gráficos de la Nación, 1962.
Estados Unidos Mexicanos. Cámara de Diputados del Congreso de los Estados Unidos Mexicanos. *Diario de los debates.* México, D.F., 1958-1967.
———. Cámara de Diputados. *Por el camino de un México nuevo-Origen, significado y perspectivas de la Constitución de 1917.* Edición de la XLVI Legislatura de la Cámara de Diputados. México, D.F., 1967.
———. Cámara de Diputados. *Los presidentes de México ante la nación: Informes, manifiestos y documentos de 1921 a 1966.* Vol. 4 Editado por XLVI Legislatura de la Cámara de Diputados. México, D.F., 1966.
———. Cámara de Diputados. *Reglamento para el gobierno interior del Congreso General de los Estados Unidos Mexicanos.* México, D.F., 1968.
———. Cámara de Senadores del Congreso de los Estados Unidos Mexicanos. *Diario de los debates.* México, D.F., 1958-1967.

BIBLIOGRAPHY

————. Comisión Nacional para el Reparto de Utilidades. *Memoria de la Primera Comisión.* 3 vols. México, D.F., 1964.

————. Secretaría de Economía. Dirección General de Estadística. *Tercer Censo Agrícola Ganadero y Ejidal, 1950, Resumen General.* México, 1956.

————. Secretaría de Industria y Comercio. Dirección General de Estadística. *Anuario estadístico de los Estados Unidos Mexicanos, 1964-1965.* México, 1967.

————. Secretaría del Trabajo y Previsión Social. *Anteproyecto, Ley Federal del Trabajo.* México, 1968.

————. Secretaría del Trabajo y Previsión Social. *Memoria del Congreso Mexicano de Derecho del Trabajo y Previsión Social, 19 al 23 de julio de 1949.* Vol. 1. México, D.F.: Talleres Gráficos de la Nación, 1950.

————. Secretaría del Trabajo y Previsión Social. *Memoria de Labores, enero a diciembre de 1964.* México, D.F.: Talleres Gráficos de la Nación, 1965.

————. Secretaría del Trabajo y Previsión Social. *Revista Mexicana del Trabajo,* vol. 10 (April 1963), special number.

López Mateos, Adolfo. *Pensamiento en acción.* Vol. 2. México, D.F.: Ediciones de la Oficina de Prensa de la Presidencia de la República, 1964.

————. *Pensamiento y programa.* México, D.F. Editorial La Justicia, 1961. Nacional Financiera, S.A. *La economía en cifras.* México, D.F., 1965.

Books

Aguilar M., Alonso, and Carmona, Fernando. *México: Riqueza y miseria.* México: Editorial Nuestro Tiempo, 1967.

Alba, Victor. *Historia del movimiento obrero en América Latina.* México, D.F.: Libreros Mexicanos Unidos, 1964.

Alexander, Robert J. *El movimiento obrero en América Latina,* México, D.F.: Editorial Roble, 1967.

Almond, Gabriel A., and Verba, Sidney. *The Civic Culture.* Boston: Little, Brown and Company, 1965.

Anderson, Charles W. *The Political Economy of Modern Spain: Policy-Making in an Authoritarian System.* Madison: University of Wisconsin Press, 1970.

Arraiza, Luis. *Historia del movimiento obrero mexicano.* México: Editorial Cuauhtémoc, 1964.

Ashby, Joe C. *Organized Labor and the Mexican Revolution under Lázaro Cárdenas.* Chapel Hill: University of North Carolina Press, 1967.

Ávila, Manuel. *Tradition and Growth: A Study of Four Mexican Villages.* Chicago: University of Chicago Press, 1969.

Bachrach, Peter, and Baratz, Morton S. *Power and Poverty: Theory and Practice.* London, New York, Toronto: Oxford University Press, 1970.

Beals, Ralph L. *Cherán: A Sierra Tarascan Village.* Washington, D.C.: Smithsonian Institution, U.S. Government Printing Office, 1946.

Bendix, Reinhard. *Max Weber: An Intellectual Portrait.* New York: Doubleday and Company, 1962.

Benveniste, Guy. *Bureaucracy and National Planning: A Sociological Case Study in Mexico.* New York: Praeger Publishers, 1970.

Brandenburg, Frank. *The Making of Modern Mexico.* Englewood Cliffs, N.J.: Prentice-Hall, 1964.

Carmona, Fernando, et al. *El milagro mexicano.* México, D.F.: Editorial Nuestro Tiempo, 1971.

Clark, Marjorie Ruth. *Organized Labor in Mexico.* Chapel Hill: University of North Carolina Press, 1934.

Cline, Howard F. *Mexico—Revolution to Evolution, 1940-1960.* London: Oxford University Press, 1962.

―――. *The United States and Mexico.* New York: Atheneum, 1963.

Cosío Villegas, Daniel, et al. *Historia moderna de México.* 7 vols. México D.F.: Editorial Hermes, 1955-1957.

Cumberland, Charles C. *Mexico: The Struggle for Modernity.* London: Oxford University Press, 1968.

D'Antonio, William, and Form, William H. *Influentials in Two Border Cities: A Study in Community Decision-Making.* South Bend, Ind.: University of Notre Dame Press, 1965.

de la Cueva, Mario. *Derecho mexicano del trabajo.* Vol. 1. México, D.F.: Editorial Porrúa, 1961.

Delhumeau Arrecillas, Antonio, et al. *México: Realidad política de sus partidos.* México: Instituto Mexicano de Estudios Políticos, 1970.

Edelman, Murray. *The Symbolic Uses of Politics.* Urbana: University of Illinois Press, 1967.

Ezcurdia, Mario. *Análisis teórico del Partido Revolucionario Institucional.* México: B. Costa-Amic, 1968.

Fagen, Richard R., and Tuohy, William S. *Politics and Privileges in a Mexican City.* Stanford: Stanford University Press, 1972.

Foster, George M. *Tzintzuntzan: Mexican Peasants in a Changing World.* Boston: Little, Brown and Company, 1967.

Fuentes Díaz, Vicente, *Los partidos políticos en México.* 2 vols. México, 1956.

Gibson, Charles. *Spain in America.* New York: Harper & Row, 1966.

Glade, William P., Jr., and Anderson, Charles W. *The Political Economy of Mexico.* Madison: University of Wisconsin Press, 1968.

BIBLIOGRAPHY

González Casanova, Pablo. *La democracia en México*. México: Ediciones Era, 1965.

Greenberg, Martin Harry. *Bureaucracy and Development: A Mexican Case Study*. Lexington, D. C. Heath and Company, 1970.

Gumpel, Henry J., and Margáin, Hugo B. *Taxation in Mexico*. Boston: Little, Brown and Company, 1957.

Hansen, Roger D. *The Politics of Mexican Development*. Baltimore: Johns Hopkins Press, 1971.

Hirschman, Albert O. *Journeys Toward Progress*. New York: Twentieth Century Fund, 1963.

Huntington, Samuel P. *Political Order in Changing Societies*. New Haven: Yale University Press, 1968.

Johnson, John J. *Political Change in Latin America: The Emergence of the Middle Sectors*. Stanford: Stanford University Press, 1958.

Johnson, Kenneth F. *Mexican Democracy: A Critical View*. Boston: Allyn and Bacon, 1971.

Kling, Merle. *A Mexican Interest Group in Action*. Englewood Cliffs, N.J.: Prentice-Hall, 1961.

Koehler, John E. *Economic Policy-Making with Limited Information: The Process of Macro-Control in Mexico*. Santa Monica, Calif.: Rand Corporation, August 1968.

Landerreche Obregón, Juan. *Participación de los trabajadores en las utilidades de las empresas*. México, D.F.: Editorial Jus, 1956.

Lerdo de Tejada, Francisco. *Manual práctico de repartición de utilidades*. Buenos Aires: Bibliografía Omeba (Colección América en Letras), 1966.

Lewis, Oscar. *Life in a Mexican Village: Tepoztlán Restudied*. Urbana: University of Illinois Press, 1951.

Lieuwen, Edwin. *Mexican Militarism: The Political Rise and Fall of the Revolutionary Army, 1910-1940*. Albuquerque: The University of New Mexico Press, 1968.

Lombardo Toledano, Vincente. *Teoría y práctica del movimiento sindical mexicano*. México: Editorial del Magisterio, 1961.

López Aparicio, Alfonso. *El movimiento obrero en México*. México: Editorial Jus, 1958.

Margáin, Hugo B. *Reparto de utilidades*. México: Publicaciones Especializadas (PESA, Selección de Estudios Latinoamericanos), 1964.

McDonald, Ronald H. *Party Systems and Elections in Latin America*. Chicago: Markham Publishing Company, 1971.

Millon, Robert Paul. *Mexican Marxist—Vicente Lombardo Toledano*. Chapel Hill: University of North Carolina Press, 1966.

Mosk, Sanford A. *Industrial Revolution in Mexico*. Berkeley: University of California Press, 1950.

Navarrete, Ifigenia. *La distribución del ingreso y el desarrollo económico de México*. México: Instituto de Investigaciones Económicas, Escuela Nacional de Economía, 1960.

Needler, Martin C. *Politics and Society in Mexico*. Albuquerque: University of New Mexico Press, 1971.

O'Donnell, Guillermo A. *Modernization and Bureaucratic-Authoritarianism; Studies in South American Politics*. Berkeley: Institute of International Studies, University of California, 1973.

Padgett, L. Vincent. *The Mexican Political System*. Boston: Houghton-Mifflin Company, 1966.

Parkes, Henry Bamford. *A History of Mexico*. Boston: Houghton-Mifflin Company, 1960.

Paz, Octavio. *Posdata*. México, D.F. Siglo Veintiuno Editores, 1970.

Pérez López, Enrique, et al. *Mexico's Recent Economic Growth: The Mexican View*. Austin: University of Texas, Institute of Latin American Studies, 1967.

Reynolds, Clark W. *The Mexican Economy: Twentieth-Century Structure and Growth*. New Haven: Yale University Press, 1970.

Ronfeldt, David. *Atencingo: The Politics of Agrarian Struggle in a Mexican Ejido*. Stanford: Stanford University Press, 1973.

Ross, Stanley, ed. *Is the Mexican Revolution Dead?* New York: Alfred A. Knopf, 1966.

Sarfatti, Magali. *Spanish Bureaucratic-Patrimonialism in America*. Berkeley: Institute of International Studies, University of California, 1966.

Schmitt, Karl M. *Communism in Mexico: A Study in Political Frustration*. Austin: University of Texas Press, 1965.

Schmitter, Philippe C. *Interest Conflict and Political Change in Brazil*. Stanford: Stanford University Press, 1971.

Schmitter, Philippe C., and Haas, Ernst B. *Mexico and Latin American Economic Integration*. Berkeley: Institute of International Studies, University of California, 1964.

Scott, Robert E. *Mexican Government in Transition*. Urbana: University of Illinois Press, 1959.

Shafer, Robert Jones. *Mexican Business Organizations: History and Analysis*. Syracuse: Syracuse University Press, 1973.

———. *Mexico: Mutual Adjustment Planning*. Syracuse: Syracuse University Press, 1966.

Simpson, Eyler. *The Ejido: Mexico's Way Out*. Chapel Hill: University of North Carolina Press, 1937.

Sommerfeld, Raynard M. *Tax Reform and the Alliance for Progress*. Austin: University of Texas Press, 1966.

Tannenbaum, Frank. *Mexico: The Struggle for Peace and Bread.* New York: Alfred A. Knopf, 1950.

―――. *Peace by Revolution: An Interpretation of Mexico.* New York: Columbia University Press, 1933.

Tarrow, Sidney G. *Peasant Communism in Southern Italy.* New Haven: Yale University Press, 1967.

Taylor, Charles Lewis, and Hudson, Michael C., eds. *World Handbook of Political and Social Indicators.* 2d ed. New Haven: Yale University Press, 1972.

Trueba Urbina, Alberto. *El nuevo Artículo 123.* México: Editorial Porrúa, 1962.

Trueba Urbina, Alberto, and Trueba Barrera, Jorge. *Ley Federal del Trabajo, reformada y adicionada.* 53d ed. México, D.F.: Editorial Porrúa, 1966.

―――. *Nueva Ley Federal del Trabajo reformada.* 17th ed. México, D.F.: Editorial Porrúa, 1972.

Tucker, William P. *The Mexican Government Today.* Minneapolis: University of Minnesota Press, 1957.

Turner, Frederick C. *The Dynamic of Mexican Nationalism.* Chapel Hill: University of North Carolina Press, 1968.

Ugalde, Antonio. *Power and Conflict in a Mexican Community.* Albuquerque: University of New Mexico Press, 1970.

Valadés, José C. *El porfirismo: Historia de un régimen.* México, D.F.: Antigua Librería Robledo, de Jose Porrúa e hijos, 1941.

Vernon, Raymond. *The Dilemma of Mexico's Development.* Cambridge, Mass.: Harvard University Press, 1963.

―――. ed. *Public Policy and Private Enterprise in Mexico.* Cambridge, Mass.: Harvard University Press, 1964.

Weber, Max. *The Theory of Social and Economic Organization.* Edited by Talcott Parsons. New York: Free Press, 1964.

Wilkie, James W. *The Mexican Revolution: Federal Expenditure and Social Change since 1910.* Berkeley: University of California Press, 1967.

Articles

Ames, Barry. "Bases of Support for Mexico's Dominant Party." *The American Political Science Review* 64 (March 1970): 153-167.

Anderson, Bo, and Cockcroft, James J. "Control and Cooptation in Mexican Politics." In *Latin American Radicalism: A Documentary Report on Left and Nationalist Movements.* Edited by Irving Louis Horowitz, Josué de Castro, and John Gerassi. New York: Random House, 1969. pp. 366-389.

Anderson, Charles W. "Bankers as Revolutionaries: Politics and Development Banking in Mexico." In *The Political Economy of Mexico*. Edited by William P. Glade, Jr., and Charles W. Anderson. Madison: University of Wisconsin Press, 1968. pp. 103-191.

Borah, Woodrow. "Colonial Institutions and Contemporary Latin America: Political and Economic Life." In *Readings in Latin American History*. Edited by Lewis Hanke. Vol. 2. New York: Thomas Y. Crowell Company, 1966. pp. 18-25.

Brandenburg, Frank. "A Contribution to the Theory of Entrepreneurship and Economic Development: The Case of Mexico." *Inter-American Economic Affairs* 3 (1962): 3-23

———. "Organized Business in Mexico." *Inter-American Economic Affairs* 12 (Winter 1958): 26-50.

———. "The Relevance of Mexican Experience to Latin American Development." *Orbis* 9 (Spring 1965): 190-213.

Camp, Roderic Ai. "The Cabinet and the Técnico in Mexico and the United States." *Journal of Comparative Administration* 3 (August 1971): 188-214.

Chalmers, Douglas A. "Political Groups and Authority in Brazil: Some Continuities in a Decade of Confusion and Change." In *Brazil in the Sixties*. Edited by Riordan Roett. Nashville: Vanderbilt University Press, 1972. pp. 51-76.

Cornelius, Wayne A. "The Impact of Governmental Performance on Political Attitudes and Behavior: The Case of the Urban Poor in Mexico City." In *Latin American Urban Research*. Vol. 3. Edited by Francine E. Rabinovitz and Felicity M. Trueblood. Beverly Hills, Calif.: Sage Publications, 1973.

———. "Nation Building, Participation and Distribution: The Politics of Social Reform under Cárdenas." In *Crisis, Choice and Change: Historical Studies of Political Development*. Edited by Gabriel A. Almond, Scott C. Flanagan, and Robert J. Mundt. Boston: Little, Brown and Company, 1973. pp. 392-498.

———. "Urbanization as an Agent of Latin American Political Instability: The Case of Mexico." *The American Political Science Review* 63 (September 1969): 833-857.

de la Cueva, Mario. "El sistema mexicano para la participación de los trabajadores en las utilidades de las empresas." *El Día*, May 4, 1968, p. 10.

"El fisco y la participación de utilidades." 20 *Investigación Fiscal* (August 1967), pp. 7-8.

Foster, George M. "The Dyadic Contract: A Model for the Social Structure of a Mexican Peasant Village." In *Peasant Society: A Reader*. Edited by Jack M. Potter, May N. Díaz, and George M. Foster. Boston: Little, Brown and Company, 1967. pp. 213-230.

BIBLIOGRAPHY

Fried, Robert C. "Mexico City." In *Great Cities of the World: Their Government, Politics, and Planning*. Edited by William A. Robson. Beverly Hills, Calif.: Sage Publications, 1967.

Friedrich, Paul. "The Legitimacy of a Cacique." In *Local-Level Politics*. Edited by Marc J. Swartz. Chicago: Aldine Publishing Company, 1968. pp. 243-269.

Fuentes Díaz, Vicente. "Desarrollo y evolución del movimiento obrero a partir de 1929." *Revista de Ciencias Políticas y Sociales* 17 (July-September 1959): 325-248.

Furtak, Robert K. "El Partido Revolucionario Institucional: Integración nacional y movilización electoral." *Foro Internacional 9* (April-June 1969): 339-353.

González, Rigoberto. "Estudio sobre la participación de utilidades en relación a las fracciones VI y IX del Artículo 123 constitucional." *Estudios proletarios*. Vol. 1. México, D.F.: Ediciones CROM, 1958. pp. 7-58.

González Blanco, Salomón. "Reformas a las fracciones II, III, VI, IX, XXI, XXII y XXXI del incisco A del Artículo 123 constitucional." *Revista Mexicana del Trabajo* 9 (May-June 1962): 7-33.

Gónzalez Casanova, Pablo. "Mexico: The Dynamics of an Agrarian and 'Semi-capitalist' Revolution." In *Latin America: Reform or Revolution?* Edited by James Petras and Maurice Zeitlin. Greenwich: Fawcett Publications, 1968. pp. 467-485.

González Navarro, Moisés. "Mexico: The Lop-Sided Revolution." In *Obstacles to Change in Latin America*. Edited by Claudio Veliz. London and New York: Oxford University Press, 1969. pp. 206-229.

Grimes, C. E., and Simmons, Charles E. P. "Bureaucracy and Political Control in Mexico: Towards an Assessment." *Public Administration Review* 29 (January-February 1969): 72-79.

Guzmán Valdivia, Isaac. "El movimiento patronal." *México: 50 años de revolución*. Edited by Julio Durán Ochoa et al. Vol. 2. *La vida social*. México: Fondo de Cultura Económica, 1961.

Hartmann, Roberto S. "Los principios de la repartición de utilidades y su aplicación en México." *Revista Mexicana del Trabajo* 9-10 (1963): 31-44.

Hernández, Octavio A.; Uruchurtu G., Alfredo; Castellanos, Jaime; Valderrama Herrera, Ernesto. "Antecedentes legales, nacionales y extranjeros" (versión definitiva). *Memoria de la Primera Comisión*. Vol. 3. México, D.F.: Comisión Nacional para el Reparto de Utilidades, 1964. pp. 617-795.

Hirschman, Albert O. "Ideologies of Economic Development in Latin America." In *Latin American Issues*. Edited by Albert O. Hirschman. New York: Twentieth Century Fund, 1961. pp. 3-36.

Huntington, Samuel P. "Social and Institutional Dynamics of One-Party Systems." In *Authoritarian Politics in Modern Society: The Dynamics of Established One-Party Systems*. Edited by Samuel P. Huntington and Clement H. Moore. New York: Basic Books, 1970.

Isbister, John. "Urban Employment and Wages in a Developing Economy: The Case of Mexico." *Economic Development and Cultural Change* 20 (October 1971): 24-46.

Izquierdo, Rafael. "Protectionism in Mexico." In *Public Policy and Private Enterprise in Mexico*. Edited by Raymond Vernon. Cambridge, Mass.: Harvard University Press, 1964. pp. 241-289.

Izquierdo, Rafael; Solís, Leopoldo M.; and Urquidi, Victor L. "La distribución del ingreso y el desarrollo económico de México." *Comercio Exterior* 11 (February 1961): 86-90.

Janos, Andrew C. "Group Politics in Communist Society: A Second Look at the Pluralistic Model." In *Authoritarian Politics in Modern Society: The Dynamics of Established One-Party Systems*. Edited by Samuel P. Huntington and Clement H. Moore. New York: Basic Books, 1970.

Jaquette, Jane S. "Revolution by Fiat: The Context of Policy Making in Peru." *The Western Political Quarterly* 25 (December 1972): 648-666.

Katz, Bernard S. "Mexican Fiscal and Subsidy Incentives for Industrial Development." *The American Journal of Economics and Sociology* 31 (October 1972): 353-358.

Linz, Juan J. "An Authoritarian Regime: Spain." In *Cleavages, Ideologies and Party Systems: Contributions to Comparative Political Sociology*. Edited by E. Allardt and Y. Littunen. Helsinki: Transactions of the Westermarck Society, 1964. pp. 291-341.

Lombardo Toledano, Vicente. "La participación de las utilidades y los intereses de la clase obrera." *Política*, vol. 4 (October 1, 1963). Supplement.

———. "The Labor Movement." *The Annals of the American Academy of Political and Social Sciences* 208 (March 1940): 48-54.

López Mateos, Adolfo. "Iniciativa de reformas a las fracciones II, III, VI, IX, XXI, XXII y XXXI del inciso 'A' del Artículo 123 de la Constitución General de la República," *Revista Mexicana del Trabajo*, special number, (April 1963), pp. 11-13.

Lowi, Theodore. "American Business, Public Policy, Case Studies and Political Theory." *World Politics* 16 (July 1964): 677-715.

Malloy, James M., "Authoritarianism, Corporatism and Mobilization in Peru." *The Review of Politics* 36 (January 1974): 52-84.

Margáin, Hugo B. "El sistema tributario." In *México: 50 años de revolución*. Vol. 1 *La economía*. Edited by Enrique Beltrán et al. México: Fondo de Cultura Económica, 1960-1962. pp. 537-567.

BIBLIOGRAPHY

McCoy, Terry L. "A Paradigmatic Analysis of Mexican Population Policy." In *The Dynamics of Population Policy in Latin America.* Edited by Terry L. McCoy. Cambridge, Mass.: Ballinger Publishing Company, 1974. pp. 377-408.

Mecham, J. Lloyd. "Mexican Federalism—Fact or Fiction?" *The Annals of the American Academy of Political and Social Sciences* 208 (March 1940): 23-38.

Menges, Constantine C. "Public Policy and Organized Business in Chile: A Preliminary Analysis." *Journal of International Affairs* 2 (1966): 343-365.

Mesa-Lago, Carmelo. "Los planes obligatorios de participación obrera en los beneficios y el desarrollo económico en Latinoamérica." *Revista Mexicana del Trabajo* 15 (October-December 1968): 185-203.

Needleman, Carolyn, and Needleman, Martin. "Who Rules Mexico? A Critique of Some Current Views of the Mexican Political Process." *Journal of Politics* 31 (November 1969): 1011-1034.

Needler, Martin C. "Changing the Guard in Mexico." *Current History* 48 (January 1965): 26-30, 52.

———. "Political Aspects of Urbanization in Mexico." In *City and Country in the Third World: Issues in the Modernization of Latin America.* Edited by Arthur J. Field. Cambridge, Mass.: Schenkman Publishing Co., 1970. pp. 287-299.

Padgett, L. Vincent. "Mexico's One-Party System: A Reevaluation." *The American Political Science Review* 51 (December 1957): 995-1008.

Pazos, Felipe. "Veinte años de pensamiento económico en la América Latina." *El Trimestre Económico* 20 (October-December 1953): 552-571.

Phelan, John. "Authority and Flexibility in the Spanish Imperial Bureaucracy." *Administrative Science Quarterly* 5 (1960): 47-65.

Pichardo, Ignacio. "Algunas consideraciones generales sobre la participación de utilidades." *Comercio Exterior* 14 (January 1964): 11-13.

Powell, John Duncan. "Peasant Society and Clientelist Politics." *The American Political Science Review* 64 (June 1970): 411-425.

Prebisch, Raúl. "El desarrollo económico de la América Latina y algunos de sus principales problemas." *El Trimestre Económico* 16 (July-September 1949): 347-431.

"El PRI." *Mañana* 978 (May 26, 1962): 31-45.

Purcell, John F. H., and Purcell, Susan Kaufman, "Machine Politics and Socio-Economic Change in Mexico." In *Contemporary Mexico: Papers of the IV International Congress of Mexican History.* Edited by James W. Wilkie, Michael C. Meyer, and Edna Monzón de Wilkie. Berkeley, Los Angeles, London: University of California Press, 1975.

Purcell, Susan Kaufman. "Authoritarianism." *Comparative Politics* 5 (January 1973): 301-312.

———. "Decision-Making in an Authoritarian Regime: Theoretical Implications from a Mexican Case Study." *World Politics* 26 (October 1973): 28-54.

Purcell, Susan Kaufman, and Purcell, John F. H. "Community Power and Benefits from the Nation: The Case of Mexico." In *Latin American Urban Research.* Vol. 3. Edited by Francine F. Rabinovitz and Felicity M. Trueblood. Beverly Hills, Calif.: Sage Publications, 1973. pp. 49-76.

Retchkiman, Benjamin. "Distribución del ingreso." *Revista de Economía* 21 (August 15, 1958): 224-230.

Reyna, José Luis. "Desarrollo económico, distribución del poder y participación política; el caso mexicano." *Ciencias Políticas y Sociales* 13 (October-December 1967): 469-486.

Rivera Marín, Guadalupe. "Los conflictos de trabajo en México, 1937-1950." *El Trimestre Económico* 22 (April-June 1955): 181-208.

Rogowski, Ronald, and Wasserspring, Lois. *Does Political Development Exist? Corporatism in Old and New Societies.* Sage Professional Papers in Comparative Politics. Beverly Hills, Calif.: Sage Publications, 1971.

Ronfeldt, David F. "The Mexican Army and Political Order since 1940." In *Contemporary Mexico: Papers of the IV International Congress of Mexican History.* Edited by James W. Wilkie, Michael C. Meyer, and Edna Monzón de Wilkie. Berkeley, Los Angeles, London: University of California Press, 1975.

Rostro, Francisco. "Estabilidad, desarrollo y sector privado." *El Día.* August 20, 1967.

Roth, Guenther. "Personal Rulership, Patrimonialism and Empire-Building in the New States." *World Politics* 20 (January 1968): 194-206.

Rottenberg, Simon. "México: Trabajo y desarrollo económico." *Foro Internacional* 11 (July-September 1959).

Schmitter, Philippe C. "Paths to Political Development in Latin America." In "Changing Latin America: New Interpretations of Its Politics and Society." Edited by Douglas A. Chalmers. *Proceedings of the Academy of Political Science* 30 (August 1972): 83-108.

―――. "Still the Century of Corporatism?" *The Review of Politics* 36 (January 1974): 85-131.

Scott, Robert E. "Legislatures and Legislation." In *Government and Politics in Latin America.* Edited by Harold Eugene Davis. New York: Ronald Press, 1958. pp. 290-332.

―――. "Mexico: The Established Revolution." In *Political Culture and Political Development.* Edited by Lucien W. Pye and Sidney Verba. Princeton: Princeton University Press, 1965. pp. 330-395.

Silvert, Kalman. "The Costs of Anti-Nationalism: Argentina." In *Expectant Peoples.* Edited by K. H. Silvert. New York: Vintage Books, 1963. pp. 347-371.

Solís, Leopoldo. "Mexican Economic Policy in the Post-War Period: The

Views of Mexican Economists." *The American Economic Review* 61, pt. 2 (June 1971): 1-67. Supplement.

Sturmthal, Adolf. "Economic Development, Income Distribution, and Capital Formation in Mexico." *The Journal of Political Economy* 63 (June 1955): 183-201.

———. "Some Reflections on Economic Development in Mexico and the Labor Movement." *Proceedings of the Seventh Annual Meeting, Industrial Relations Research Association, Detroit, Michigan, December 28-30, 1954.* pp. 60-68.

Taylor, Philip B., Jr. "The Mexican Elections of 1958: Affirmation of Authoritarianism?" *Western Political Quarterly* 13 (September 1960): 722-744.

Tuohy, William S., and Ronfeldt, David. "Political Control and the Recruitment of Middle-Level Elites in Mexico: An Example from Agrarian Politics." *Western Political Quarterly* 22 (June 1969): 365-374.

Urquidi, Victor L. "El impuesto sobre la renta en el desarollo económico en México." *El Trimestre Económico* 23 (October-December 1956): 424-437.

———. "La perspectiva del crecimiento económico y la repartición del ingreso nacional." *Comercio Exterior* 9 (April 1959): 198-203.

———. "Problemas fundamentales de la economía mexicana." *Cuadernos Americanos* 114 (January-February 1961): 69-103.

Walton, John, and Sween, Joyce A. "Urbanization, Industrialization and Voting in Mexico: A Longitudinal Analysis of Official and Opposition Party Support." *Social Science Quarterly* 52 (December 1971): 721-745.

Wiarda, Howard J. "Toward a Framework for the Study of Political Change in the Iberic-Latin Tradition: The Corporative Model." *World Politics* 25 (January 1973): 206-235.

Wilkie, James W. "New Hypotheses for Statistical Research in Recent Mexican History." *Latin American Research Review* 6 (Summer 1971): 3-17.

Wionczek, Miguel S. "Incomplete Formal Planning: Mexico." In *Planning and Economic Development.* Edited by Everett Hagen. Homewood, Ill.: Irwin Press, 1963. pp. 150-182.

Wolf, Eric, "Aspects of Group Relations in a Complex Society: Mexico." In *Contemporary Cultures and Societies of Latin America.* Edited by Dwight B. Heath and Richard N. Adams. New York: Random House, 1965. pp. 85-101.

———. "Types of Latin American Peasantry: A Preliminary Discussion." *American Anthropologist* 3, pt. 1 (June 1955): 452-471.

Yllanes Ramos, Fernando. "La ley mexicana sobre participación de utilidades a los trabajadores." *Revista Mexicana del Trabajo* 11 (May-June 1964): 21-44.

Pamphlets

Confederación de Cámaras Nacionales de Comercio (CONCANACO). *XLV Asamblea General Ordinaria.* México, D.F., September 1962.

Confederación de Trabajadores Mexicanos (CTM). Comité Nacional. *Informe al XLV Consejo Nacional Ordinario, 29, 30, 31 de julio, 1952.* México, D.F.

―――. *Convocatoria a la Reunión Nacional del Trabajo y Seguridad Social, 11-12 de mayo, 1965.*

―――. *Informe del Comité Nacional: LXVII Asamblea General Ordinaria del Consejo Nacional, agosto 30, 31, septiembre 1, 2 de 1963.* México, D.F.

Confederación Patronal de la República Mexicana (COPARMEX). *Noticia histórica sobre el Departamento para el Estudio de la Participación de Utilidades y los Salarios Mínimos.* México, D.F.

―――. *La participación en las utilidades: Estudio de CONCAMIN, CONCANACO y COPARMEX.* Serie Documentos y Discursos, No. 3 México, D.F., July 3, 1953.

―――. Departamento de Relaciones Públicas, Servicio de Prensa. Boletín #SPCP-60/64. May 2, 1964.

―――. Departamento para el Estudio de la Participación de Utilidades y Salarios Mínimos. *Estudio sobre la participación de utilidades.* 11 vols. México, 1963.

―――. Departamento para el Estudio de la Participación de Utilidades y Salarios Mínimos. *Estudio sobre los sistemas de aplicación de utilidades en favor de los trabajadores.* México, D.F., September 1963.

Confederación Regional de Obreros Mexicanos (CROM). *Memoria de la CROM, 1961-1963.* México, D.F., 1963.

―――. *Memoria de los trabajos realizados por el Comité Central durante su ejercicio del 1 de agosto de 1951 al 31 de julio de 1953.* México, D.F., 1953.

―――. *Memoria de los trabajos realizados por el H. Comité Central durante su ejercicio del 1 de agosto de 1955 al 31 de julio de 1957.* México D.F., 1957.

Partido de Acción Nacional. *Plataforma política de Acción Nacional (Aprobada en la XIII Convención Nacional del Partido—24 noviembre de 1957).*

―――. *Plataforma que sostendrá el PAN en la campaña electoral para renovación de poderes federales en 1952 y que fué aprobada por la Convención Nacional reunida en la Ciudad de México del 17 al 20 de noviembre de 1951.* México, D.F.

Partido Revolucionario Institucional. *Declaración de principios.* México, D.F., 1966.

―――. "Dictamen sobre 'La Declaración de Principios,' III Asamblea

BIBLIOGRAPHY

Nacional del PRI, March 27, 1960, PRI archives, Mexico City.
――――. *Estatutos*. México, D.F., 1966.
――――. *El PRI, ideología y composición*. México, D.F., 1965.

Unpublished Materials

Adie, Robert Frank. "Agrarianism in the Mexican Political System."
Ph.D. dissertation, University of Texas, Austin, 1970.

Alvírez Friscione, Alfonso. "Fundamentos axiológico-jurídicos de la
participación en las utilidades y efectos sociológico-jurídicos de esta
institución." Tesis profesional, Universidad Nacional Autónoma de
México, Facultad de Derecho. México, D.F., 1965.

Anderson, Roger Charles. "The Functional Role of the Governors: Their
States in the Political Development of Mexico, 1940-1964." Ph.D.
dissertation, University of Wisconsin, 1971.

Austin, R. V. "The Development of Economic Policy in Mexico with
Special Reference to Economic Doctrines." Ph.D. dissertation, University of Iowa, 1958.

Brandenburg, Frank Ralph. "Mexico: An Experiment in One-Party
Democracy." Ph.D. dissertation, University of Pennsylvania, 1956.

Brown, Lyle C. "General Lázaro Cárdenas and Mexican Presidential
Politics, 1933-1940: A Study of the Acquisition and Manipulation of
Political Power." Ph.D. dissertation, University of Texas, 1964.

Camp, Roderic Ai. "The Role of the *Técnico* in Policy-Making in Mexico:
A Comparative Study of a Developing Bureaucracy." Ph.D. dissertation, University of Arizona, 1970.

Chalmers, Douglas A. "Parties and Society in Latin America." Paper
prepared for delivery at the 1968 Annual Meeting of the American
Political Science Association. Washington, D.C. Mimeographed.

Coleman, Kenneth M. and Wanat, John. "Models of Political Influence in
Federal Budgetary Allocations to Mexican States." Paper prepared for
delivery at the 1973 Annual Meeting of the American Political Science
Association, New Orleans. Mimeographed.

Creagan, James Francis. "Minority Political Parties in Mexico: Their Role
in a One-Party Dominant System." Ph.D. dissertation, University of
Virginia, 1965.

Eckstein, Susan Eva. "The Poverty of Revolution." Ph.D. dissertation,
Columbia University, 1972.

Everett, Michael David. "The Role of the Mexican Trade Unions,
1950-1963." Ph.D. dissertation, Washington University, Missouri, 1967.

Gabbert, Jack. "The Evolution of the Mexican Presidency." Ph.D. dissertation, University of Texas, 1963.

Guerrero, Euquerio. "Participación de los trabajadores en las utilidades de las empresas." February 1961. Mimeographed.

Kaufman, Robert R. "Corporatism, Clientelism, and Partisan Conflict in Latin America: A Comparative Study of Argentina, Brazil, Chile, Colombia, Mexico, Venezuela and Uruguay." Manuscript, 1973.

Linz, Juan J. "Notes toward a Typology of Authoritarian Regimes." Paper prepared for delivery at the 1972 Annual Meeting of the American Political Science Association, Washington D.C. Mimeographed.

Margiotta, Franklin D. "Changing Patterns of Political Influence: The Mexican Military and Politics." Paper prepared for delivery at the 1973 Annual Meeting of the American Political Science Association, New Orleans. Mimeographed.

Menges, Constantine Christopher. "The Politics of Agrarian Reform in Chile: The Role of Political Parties and Organized Interest Groups." Ph.D. dissertation, Columbia University, 1968.

Meyers, Frederic. "Party, Government and the Labor Movement in Mexico: Two Case Studies." Paper prepared for the International Institute for Labour Studies Research Conference on Industrial Relations and Economic Development, Geneva, Switzerland, 1964. Mimeographed.

Michaels, Albert Louis. "Mexican Politics and Nationalism from Calles to Cárdenas." Ph.D. dissertation, University of Pennsylvania, 1966.

Miller, Richard Ulric. "The Role of Labor Organizations in a Developing Country: The Case of Mexico." Ph.D. dissertation, Cornell University, 1964.

Mirin, Linda S. "Public Investment in Aguascalientes: A Study of the Politics of Economic Policy." Ph.D. dissertation, Harvard University, 1964.

Mirin, Linda S., and Stinchcombe, Arthur L. "The Political Mobilization of Mexican Peasants." Mimeographed. The Johns Hopkins University.

Mundale, Charles Irving. "Local Politics, Integration and National Stability in Mexico." Ph.D. dissertation, University of Minnesota, 1971.

Padgett, Leon Vincent. "Popular Participation in the Mexican 'One-Party' System" Ph.D. dissertation, Northwestern University, 1955.

Partido Revolucionario Institucional (PRI). "III Asamblea Nacional del PRI, 1960." Folder. PRI Archives, Mexico City.

Purcell, John F. H. and Purcell, Susan Kaufman. "Mexican Business and Public Policy." Paper presented at a Conference on Authoritarianism and Corporatism in Latin America, April 4-6, 1974, University of Pittsburgh.

207

BIBLIOGRAPHY

Rabinovitz, Francine. "Decision-Making for Development in Mexico City." Mimeographed.
Reyes Ponce, Agustín. "Estudio sobre la participación legal de los trabajadores en las utilidades de las empresas," June 9, 1962. Mimeographed.
Richmond, Patricia McIntire. "Mexico: A Case Study of One-Party Politics." Ph.D. dissertation, University of California, Berkeley, 1965.
Schers, David. "The Popular Sector of the Mexican PRI." Ph.D. dissertation, University of New Mexico, 1972.
Tuohy, William S. "Institutionalized Revolution in a Mexican City." Ph.D. dissertation, Stanford University, 1967.
Ugalde, Antonio; Olson, Leslie; Schers, David; and Von Hoegen, Michael. "The Urbanization Process of a Poor Mexican Neighborhood: The Case of San Felipe del Real Adicional, Juárez." Mimeographed. 1973.
Von Sauer, Franz Alfred. "Ideological Politics in Mexico and the Partido Acción Nacional: A Case Study in Political Alienation." Ph.D. dissertation, Georgetown University, 1971.

Newspapers

Ceteme (1950-1972), Mexico City
El Día (1962-1972), Mexico City
El Nacional (1950-1972), Mexico City
El Universal (1950-1972), Mexico City
Engrane (1965-1968), Mexico City
Excelsior (1950-1972), Mexico City
La Prensa (1967-1968), Mexico City
The New York Times (1950-1974), New York City

Magazines

Boletín Quincenal (1958-1974), Mexico City
La Nación (1960-1962), Mexico City
La República (1960-1968), Mexico City
Mexican-American Review (1963-1972), Mexico City
Política (1960-1968), Mexico City
Revista Bancaria (1960-1968), Mexico City
Siempre! (1961-1968), Mexico City
The Economist para América Latina (1969-1970), London
Voz Patronal (1960-1968), Mexico City

Index

See also Labor, Ministry of
Labor law. See Federal Labor Law
Labor movement, 15, 20-26, 28, 43,
50, 51, 109, 135, 137, 143; as sector
of PRI, 9, 18, 35, 93; and profit
sharing, 9, 48, 49-58, 62-63, 67, 71,
74, 75, 77, 80, 81-82, 83, 90-91, 104,
108, 109, 112, 114, 119-129 passim,
132, 135-136, 140, 142, 144-145,
146; economic position of, 22, 68;
leaders of, 22-24, 38, 50, 66, 80, 90,
104, 109; disunity in, 23, 24, 54-56,
57-58, 62-63; and regime, 23-24, 28,
29, 38, 44, 67, 76, 84, 98-99, 104,
113, 114, 116, 142, 144-145, 147;
and private sector, 24, 63-64, 71-72,
78, 82, 91-92, 104, 119, 122, 124-
125, 144; compared with private
sector, 28, 116; and nonunionized
workers, 35, 68; and rural workers,
35, 144; represented on Profit-Sharing Commission, 47, 98, 99, 106,
108-109, 143-144; revolutionary
stage of, 49-50; becomes more moderate, 50, 60, 62; mobilizes rank and
file, 65-66, 76, 84, 91-92
LAFTA. See Latin American Free
Trade Association
Landerreche Obregón, Juan, 61, 105
Landholdings. See Agrarian interests;
Ejido system; Peasants
Latin American Free Trade Association (LAFTA), 130; decision, 131-
136 passim
Lebrija, Rafael, 86, 97
Legislature. See Chamber of Deputies;
Congress; Senate
Leyva Velázquez, General, 58-59
Linz, Juan, 5
Lombardo Toledano, Vicente, 21, 24,
105
López, Jacinto, 19
López, Prudencio, 128
López Mateos, Adolfo, 11, 67, 72, 77;
and profit-sharing decision, 9, 47,
49, 51, 56, 57-58, 60, 62, 64-65, 67-
68, 69, 70, 71, 73-75, 76, 78, 79-80,

81, 83-84, 90, 101, 103-104, 123,
133, 136; and organized labor, 24,
25, 67, 69, 90, 104, 109; and
creation of CNT, 25; and strikes,
25, 66; and PRI, 44, 62; and PAN,
62; prolabor stance of, 65, 66-67,
69, 72, 113, 145-146; as secretary of
labor, 65, 68, 133; and Federal
Labor Law reforms, 65, 81, 83-84,
90; liberal image of, 67, 79; and
Profit-Sharing Commission, 74, 94,
95, 103-104, 113, 115-116, 122; and
private sector, 78-80
Lower classes, 1, 2, 7-8, 69. See also
Peasants; Urban groups

McCoy, Terry L. 130, 134
Madariaga, Sánchez, 23
Madrazo, Carlos, 38-39
Margáin, Hugo P., 73, 102, 115, 118;
as head of Tax Bureau, 70; as
technocrat, 94-95, 141; as head of
Profit-Sharing Commission, 94-96,
100, 103, 105, 106-107, 109, 110-
112, 113, 114, 119, 120-121, 122;
and labor, 103, 106-107, 109, 110,
119; and private sector, 103, 106-
107, 110-112, 119, 120-121; on
"prior deductions," 106-107, 110,
112; and business-labor compromise, 110, 113
Middle-sector groups, 2, 16, 18, 26,
27, 147. See also CNOP; PRI
Miners' Union, 54, 58, 99. See also
STMMRM
Minimum wage, 47, 48, 59, 65-66, 77,
89, 113, 126
Mobilization, 12, 13, 14-15, 17, 32,
34, 143, 146, 147; regime-sponsored, 31, 32, 75-82, 103-104, 135-
136, 139; level of, 31, 34-38, 45, 131,
140; during elections, 31, 35, 36;
and rural population, 32, 35; and
PRI, 36, 42; discouraged by regime,
74, 75-82, 103-104, 137, 138, 141-
142; and profit sharing, 75-82, 103-
104; of rank and file of business, 84,